KINETICS OF DIFFUSION CONTROLLED CHEMICAL PROCESSES

Institute of Chemical Physics
USSR Academy of Sciences

KINETICS OF DIFFUSION CONTROLLED CHEMICAL PROCESSES

A.A. Ovchinnikov,
S.F. Timashev, and A.A. Belyy

NOVA SCIENCE PUBLISHERS
Commack

Nova Science Publishers, Inc.
283 Commack Road
Suite 300
Commack, New York 11725

This book is being published under exclusive English language rights granted to Nova Science Publishers, Inc. by the All-Union Copyright Agency of the USSR (VAAP).

Library of Congress Cataloging-in-Publication Data
available upon request

ISBN 0-941743-52-7

The original Russian-language version of this book was published by Khimiya Publishing House in 1986.

Printed in the United States of America

CONTENTS

FOUNDATIONS OF DIFFUSION CONTROLLED REACTION KINETICS

DIFFUSION CONTROLLED PROCESSES WITH INTERMOLECULAR INTERACTION

METHODS OF MANY-PARTICLE DESCRIPTION IN THE KINETICS OF DIFFUSION CONTROLLED PROCESSES

BOUNDARY CONDITIONS IN DIFFUSION CONTROLLED REACTION KINETICS

MEMBRANE SEPARATION PROCESSES FOR LIQUID AND GAS MEDIA

CHAPTER 1

FOUNDATIONS OF DIFFUSION-CONTROLLED REACTION KINETICS

1.1 BASIC PRINCIPLES OF THE KINETICS OF DIFFUSION CONTROLLED REACTIONS

Diffusion kinetics studies processes including as one of the stages the transport of components which can be described by diffusion equations. If these processes are accompanied by a chemical reaction, then we usually speak about the kinetics of diffusion controlled reactions (DCR). In DCR a special role is played by reactions in which the elementary act of chemical transformation is limited by the transport of reacting particles to each other. The reacting particles may be of various origin—ions, electrons, lattice defects, quasi-particles, dislocations, etc. To avoid concrete definitions of the object, we shall use the term active particles. The specific features of DCR kinetics can be clearly seen in comparison to classical chemical kinetics.

Classical chemical kinetics of gas-phase reactions postulates the existence of a quasi-equilibrium in the system with constant temperature T in space and time, and a uniform reagent concentration (or P pressure) throughout the volume of the system [14,15]. The rate of the chemical reaction is supposed to be sufficiently low so that the uniform distribution of particles is not disturbed. Thus it is postulated that the rate of the reaction is a function of the system state, completely determined by the values of P, T and the reagent concentration [15]. Such assumptions are substantiated only when diffusion space relaxation occurs sufficiently quickly so that diffusion does not control the overall reaction rate.

If the rate of diffusion relaxation is sufficiently low (or the rate of the chemical reaction high) the distribution of particles in the system becomes inhomogeneous. Generally this means that the state of the system must be determined not only by the average concentration of reagents but also by the relative positions and velocities of particles. Moreover, since the linear dimensions of the inhomogeneities can be compared to the mean distance between molecules, the notion of particle concentration becomes meaningless in the general case and the state of the system must be described by the multiparticle distribution function [16].

Such a situation complicates considerably the search for the probability of particle collision (or the number of collisions). It should also be remembered that the course of the chemical reaction itself determines mutual correlation in the positions of the particles. Thus the probability of a collision between two particles becomes a complex function of the coordinator of all the particles in the system, and thus we come to the many-body problem. In working out this problem, methods of quantum mechanics and statistical physics are receiving wide application. In this particular sense, the development of the DCR theory can be divided into two stages. In the first stage, simple models based on single-particle distribution functions were worked out (the role of particle correlation had been ignored). In the second, a new stage—the role of correlation effects—was revealed raising multiparticle distribution functions.

In formulating different approaches to the description of DCR kinetics, a number of conjectures are being made, depending on the specific character of the problem and the extent of generality in the statement of the problem.

Let us enumerate the basic assumptions connected with the possibility of offering a diffusion description of the dynamics of chemically active particles:

1. The system containing chemically active "primary" reagents in the state of thermal equilibrium [17]. Processes studied by the DCR theory are as a rule rather complicated because of the variety of the proceeding reactions. These reactions include "fast" and "slow" stages. Chemical transformations passing at fast stages are accompanied by a relaxation of the system to thermal equilibrium and create the initial product for "slow" chemical stages. The kinetics of the latter ones are studied by the DCR theory.

Radiation chemical processes serve as a perfect example of such reactions [18]. The passing of high-energy radiation through a substance gives rise to various types of active particles: excited molecules and ions, radicals, etc. The resultant active particles are non-uniformly distributed in space and compose either tracks or cords, the structure of which depends on both the type and energy of the ionizing radiation and the state and structure of the absorbing substance. On exposure to radiation, initial periods of relaxation of the system are characterized by the attainment of a thermal equilibrium rather than a chemical one. Thermal relaxation creates stable molecules, often different from the initial ones, and chemically unstable products. Subsequent processes determined by the chemical transformations of these unstable active particles and their possible diffusion lead to establishing a chemical (and with it a thermodynamic) equilibrium.

2. At the initial time, chemically active particles are distributed inhomogeneously in the general case.

3. Active particles diffuse according to the laws of macroscopic diffusion, i.e., Fick's laws are applicable. Essentially, this is the main postulate of the DCR theory which however needs to be further discussed and substantiated. The applicability of this approach is undeniable in the absence of strong macroscopic concentration gradients. But in case of spatial-inhomogeneous particle distribution, the notion of concentration can become meaningless. A typical example is the DCR elementary act,

which is impossible to consider in terms of "concentrations". In essence this points to the necessity of analyzing diffusive mobility on the microscopic molecular-kinetic level.

Having a most general idea of the structure of solid and liquid media, it can be supposed that the diffusing active particle can energetically occupy equivalent positions on the points of an imagined lattice and migrate by jumping into one of the equivalent positions within a short time interval. It is assumed that immediately after each elementary jump (transport) the particle reaches thermal equilibrium with the medium (as it follows from basic assumption 1). This is equivalent to the hypothesis on the statistical independence of such steps of particles. This assumption admits the use of the theory of stochastic processes for analyzing the average particle movement. The application of random process theory actually includes the DCR theory in statistical physics, and the use of its methods accounts for considerable progress in studying the kinetics of diffusion limited reactions.

The random walk of the active particle about the lattice points after a large number of jumps (see [19]) is described by a diffusion equation. The particle flux is expressed by concentration gradients according to Fick's laws. The applicability of this approach has been widely discussed earlier [20–24].

But it must be taken into account that Fick's laws have an approximate character. Modern experimental studies quite often require stepping beyond the limits of diffusion approximation, and the methods of the random processes theory admit this possibility. In particular, description within the framework of macroscopic (Fick's) kinetics may prove inadequate for analyzing processes at physically accessible short times, and a more suitable description in terms of discrete models of random walks might be required [19]. In particular, it is possible to derive an adequate formulation of boundary conditions [25] and find methods that take into account the chemical transformation which accompanies diffusion. The complicated character of the latter problem points to the necessity of introducing a special (fourth) assumption on which the DCR theory is based.

4. The chemical reaction determines boundary conditions of diffusion equations. The simplest way of discussing this proposition is to present the system (at the beginning of the process) in a quasi-equilibrium state with a uniform distribution of components. In the absence of chemical reactions, the system will be in overall equilibrium. The chemical reactions change the state of the system. However it retains homogeneity due to the multiplicity of elementary acts and the random character of collisions. Macroscopic gradients of concentration do not arise, and the adequate taking into account of limiting transport (diffusive) stages of chemical transformation is far from evident. This problem was solved by Smoluchowski. He took note of the fact that the gradients of the reacting particle distribution function would appear due to the acts of a chemical reaction and the latter can be taken into account by boundary conditions for a diffusion equation [20,26,27]. The Smoluchowski approach envisages a search for the conditional probability for a particle to remain unreacted by the time moment considered instead of a routine search for the number of collisions per unit time in this case. The dynamics of unreacted particles is still described by a diffusion equation. If for the sake of simplicity we assume the independence of the random walks of separate particles

and ignore the collisions between particles of the same type. It is possible to consider the interaction dynamics of a pair of the different particles conventionally locating one of them at the origin of the coordinates. Then calculating the conditional probability of the "survival" of particles is reduced to considering a diffusion equation where the surface of the immovable particle will play the role of a boundary. If we assume the condition on the surface of the immovable particle assigned and assume the initial distribution given, it becomes possible to find the corresponding probability of the survival of a pair of particles. The probability of the reaction (or the number of collisions) will then be determined by a flux of mobile particles through the boundary assigned by the selected particle surface. Thus taking into account the chemical transformations in DCR kinetics is reduced to the problem of boundary conditions.

1.2 THE SMOLUCHOWSKI METHOD

The DCR theory was first developed by Smoluchowski for solving problems of the kinetics of coagulation of colloids [26]. The method describing the kinetics of diffusion controlled reactions suggested in [26] was called the Smoluchowski method, or the concentration gradient method. As has already been mentioned, this method promotes the idea of accounting for the chemical reaction by introducing the corresponding boundary condition for a diffusion equation. The method of doing this and the establishment of the physical sense of the implicit assumption requires special analysis. This becomes even more important since in his solution Smoluchowski actually formulated two different problems which have become the classical objects of DCR theory.

Let us consider a viscous medium containing spherical particles per unit volume each particle of the same radius R_A. Let the movement of each particle be Brownian and described by a diffusion equation with diffusion coefficient D_A. Supose that the collision of two particles results in an "instantaneous" reaction, removing both particles from the considered system, i.e., a bimolecular irreversible reaction $A + A \rightarrow C$ is under study. The task is to calculate the concentration of particles that remain by time t after the start of the reaction.

To solve this problem, Smoluchowski introduces a new auxiliary problem. Let us assume that one of the particles (having radius R_A) is in a fixed position, whereas the others noncolliding in the course of their Brownian motion can disappear only when encountering the sphere of radius R_A—the surface of the selected particle. The sphere surface acts as an ideal sink (absorbing screen), i.e., the encounter of any particle with the sphere leads to immediate absorption. As for the selected particle, it can be assumed that it either quits the system upon colliding with another particle, or absorbs other particles while remaining in the system. It is the latter case that is considered by Smoluchowski. The fixed particle acts as a sink of infinite capacity (this particle denoted by B may actually be considered as a different type of particle than mobile particles A), the number of sinks being much fewer than the full number of A particles. Due to this fact, the reaction between Brownian particles A and sink particles B is monomolecular:

$$A + B \longrightarrow C \tag{1.1}$$

The corresponding problem of particle diffusion to sinks of infinite capacity can be treated macroscopically. It is then reduced to a problem on an absorbing center when the concentration of sinks is low and their "effect" upon the system of A particles can be considered independent. The usual formulation of the problem is as follows [5].

The local concentration of A particles in the immediate neighborhood of the absorbing sphere $c(r,t)$ will depend only on time t and distance r between the center of the absorbing sphere and the center of the particle. The concentration distribution of A particles can be represented by the diffusion equation

$$\partial(rc)/\partial t = D\partial^2 cr/\partial r^2 \tag{1.2}$$

with the initial condition

$$c(r,0) = c_0 \qquad \text{at} \qquad r > R \tag{1.3}$$

where R is the radius sum of the mobile and fixed particles.

Since the absorbing particle is an ideal sink the equation

$$c(R,t) = 0, \qquad t > 0 \tag{1.4}$$

can be taken for the boundary condition.

The second boundary condition

$$c(\infty,t) \equiv C(t) = c_0 \qquad \text{at} \qquad r \longrightarrow \infty \tag{1.5}$$

is inadequate for the above-formulated problem because the number of particles in the system diminishes with time. Nevertheless the consideration of such a model problem allows us to find the solution of the original problem as the reservoir of active particles is practically inexhaustible at low concentrations of traps.

The solution of equation (1.2) at given conditions (1.4)–(1.5) is easily found [5]. Most interesting is the expression for the flux $j(t)$ of particles through the absorbing sphere of radius R:

$$j(t) = 4\pi DR^2 \left.\frac{\partial c(r,t)}{\partial r}\right|_{r=R} = 4\pi DRc_0\left[1 + \frac{R}{\sqrt{\pi Dt}}\right] \tag{1.6}$$

The flux $j(t)$ is equal to the average number of particles crossing the surface of radius R per time unit at time t owing to Brownian motion. At sufficiently large time $t \gg R^2D^{-1}$, the flux assumes a stationary value

$$j(\infty) = 4\pi DRc_0 \tag{1.7}$$

corresponding to the stationary distribution of the concentration

$$c(r,\infty) \equiv C(r) = c_0[1 - (R/r)] \tag{1.8}$$

The full quantity of particles absorbed by time t after the beginning of the process is shown in the expression

$$M(t) = \int_0^t j(\tau)d\tau = 4\pi DRc_0[t + (2R\sqrt{t} / \sqrt{\pi D})] \tag{1.9}$$

The volume of the system is implied to be infinite ($V \to \infty$), as is the number of particles in the system, $N_0 \to \infty$, but $N_0 V^{-1} \to c_0$ (otherwise equation (1.6) is meaningless). However the volume of any system is always finite, the same as the number of particles it contains and the number N_s of absorbing sinks whereas an infinite number of particles will be absorbed in an infinite time period as follows from (1.9).

To avoid this difficulty (clearly indicated in [28]), Smoluchowski resorts to a probabilistic interpretation of the problem.

Let us formally assume $N_0 = 1$. Then the value $M(t)$ acquires the sense of probability ω_t that some particle located somewhere in volume V at the initial time moment will join the selected particle by time t, i.e.

$$\omega_t = 4\pi DRV^{-1}[t + (2R\sqrt{t} / \sqrt{\pi D})] \tag{1.10}$$

This necessarily requires $\omega_t \ll 1$. Now let us introduce probability μ_t of some mobile particle not being absorbed in time t: $\mu_t = 1 - \omega_t$. If now we consider volume V holding $\rho_s V$ independent sinks (ρ_s is the volume concentration of sinks), then the probability $\overline{\mu}_t$ of one particle not being absorbed by any sink by time t is

$$\overline{\mu}_t = (1 - \omega_t)^{\rho_s V}$$

Using the well-known limit relationship $\lim_{N\to\infty} (1 - N^{-1})^N = e^{-1}$ it is possible to present

$$\overline{\mu}_t \approx e^{-\omega_t \rho_s V}$$

and taking into account (1.10) the expression assumes the form

$$\overline{\mu}_t \sim \exp\{-4\pi DR\rho_s[t + (2R\sqrt{t/\pi D}]\}$$

(only now the limit $t \to \infty$ acquires meaning).

If a volume unit holds c_0 independent particles at the initial time moment, the concentration of the remaining mobile particles by time t is determined through the expression

$$C(t) = c_0 \exp\{-4\pi D\rho_s[t + 2R\sqrt{t/\pi D}]\} \tag{1.11}$$

which presents the reaction kinetics (1.1). This solves the auxiliary problem.

Now it is quite simple to find the solution of the second and main problem, i.e., to describe the kinetics of the decrease of particles of the same type in the reaction $A + A \to C$. The mobility of the sink itself must necessarily be taken into account in solving the problem. It can be easily shown that the relative movement of two particles accomplishing a Brownian motion with coefficients D_1 and D_2 is a Brownian motion characterized by the constant $D = D_1 + D_2$. For particles of the same type it should be assumed that $D = 2D_A$. The probability of the collision of two particles may be derived from (1.6) if the interpretation of the flux is probabilistic. In this case the probability of an encounter of one of the particles with the sink per time unit (which is the probability of the collision of two particles) is

$$\nu_t = j(t)/N_0 = 4\pi DRV^{-1}(1 + R/\sqrt{\pi Dt})$$

Since the probability of finding either the sink or the selected particle in an arbitrary point of volume V is simply V^{-1}, the absolute probability of the collision of two particles is $\bar{\nu}_t = \nu_t V^{-1}$. The number of pairs Z, capable of collision at time t is determined by the number N_t of particles survived at this moment

$$Z = \frac{1}{2} N_{t}2$$

so the kinetic equation of the bimolecular reaction assumes the form

$$\frac{dC(t)}{dt} = -\bar{\nu}_t Z = C^2(t)4\pi D_A R(1 + R/\sqrt{2\pi D_A t}) \equiv -k_{ex}(t)C^2(t) \qquad (1.12)$$

where $k_{ex}(t)$ is the effective constant of reaction rate.

It was for a special purpose that we have given a detailed account of Smoluchowski's original work. We wished to accentuate the intuitive heuristic character of this derivation and to point out the fully acceptable (Physically) results.

A thorough analysis of the Smoluchowski method was given by Collins and Kimball [29]. We shall briefly characterize their approach to the problem. Let us once again refer to equation (1.2) with conditions (1.3)–(1.5) and introduce a dimensionless function

$$\varphi(r,t) = 1 - c_0^{-1}c(r,t) \qquad (1.13)$$

It can be clearly seen that function $\varphi(r,t)$ satisfies the equation

$$\frac{\partial(r\varphi)}{\partial t} = D\frac{\partial^2(r\varphi)}{\partial r^2} \qquad (1.14)$$

with conditions

$$\varphi(r,0) = 0, \quad \varphi(R,t) = 1, \quad \varphi(\infty,t) = 0$$

Function $\varphi(r,t)$ (see [5]) can be considered the probability for a mobile particle (which at moment $t = 0$ is at distance r from the absorbing sphere) to reach the absorbing boundary at least once by time t on the condition that it has not been absorbed by other A particles, and the selected particle is fixed. In passing to absolute probability, the probability for a selected particle to be created at a given point of volume V must be taken into account. Then the function $\nu(r,t) = 1 - \varphi(r,t)$ that satisfies the diffusion equation with conditions $\nu(r,0) = 1$, $\nu(R,t) = 0$, $\nu(\infty,t) = 1$ have the sense of conditional probability of the particle not colliding even once with the selected particle sink in time t. It is function $\nu(r,t)$ that is analyzed by Smoluchowski up to a normalization constant on the basis of equations (1.2–1.5). Thus the essence of the Smoluchowski method is to consider the conditional probability for a particle to remain unreacted by the given time moment. The dynamics of the unreacted particle can be shown by a diffusion equation and only then can a chemical reaction be introduced with the help of a boundary condition.

This particular aspect of the Smoluchowski method was brought to light by Collins and Kimball [29].

It must be stressed however that finding the corresponding conditional probability for a pair of particles is not yet sufficient for solving both of the problems put forward by Smoluchowski. The simpler problem of diffusion in the presence of sinks of infinite capacity requires taking into account the "competition" of sinks since every mobile particle can be absorbed on any of them. It has already been shown above that an attempt to take into account the effect of preserving the notion of "concentration" leads to "badly defined" quantities ω_t or $\overline{\mu}_t$. This indicates the necessity of ruling out the "direct" use of the notion of particle concentration and of proceeding to specially designed distribution functions. In this case, taking into account correlation effects points to the necessity of making more precise the equation for the probability density. The usual diffusion equation becomes inadequate: the equation must contain additional terms related to particle correlation.

The specificity of the random walks themselves also assumes great importance. In particular, the dimensions of the corresponding space. The original two-particle problem in the Smoluchowski method may not, for example, have stationary solutions as in the two-dimensional case. And one-dimensional problems become the most difficult to solve. If the movements are strictly one-dimensional a special role is played by the so-called "trolley-bus effect" when a pair of selected particles cannot react till the particles inside the interval between them disappear. Only the disappearance of the "inside" particles permits the selected pair to approach each other at a reaction distance.

When solving one-dimensional problems, reactions in a narrow submicropore, for example, (the transversal dimension is much smaller than the length, but larger than the diameter of the molecules) it is not clear beforehand how to calculate the probability of collisions. In this case mutual "overshoots" are possible and the situation actually becomes three-dimensional. This means that equations analogous to those of formal kinetics cannot be derived for mean concentrations.

1.3 THE PROBLEM OF BOUNDARY CONDITIONS

In the DCR theory, chemical transformations are described by boundary conditions to mass-action equations. This accounts for the special role played by boundary conditions in the analysis of the reaction kinetics. The adequate formulation of boundary conditions becomes a problem in itself. The specificity of this problem is determined by the complex hierarchy of the microscopic and macroscopic approaches to the description of the general reaction kinetics. The specificity of the elementary act requires proceeding to the microscopic description of reaction systems and introducing the distribution function instead of particle concentration.

The study of the problem of boundary conditions served as a starting point for a deep analysis of the Smoluchowski method and for thus determing the conceptual base of the DCR theory.

Discussions of the Smoluchowski method along these lines were begun by Collins and Kimball [29], who formulated the basic questions. Let us reproduce their considerations using the example of the problem of sinks (monomolecular reaction), when the rate of the reaction is characterized by a flux into the absorbing sinks. According to [29] the main difficulties in the Smoluchowski method arise in the discussion of the following two principles:

1. The flux into the absorbing sphere obtained in accordance with the Smoluchowski theory is equal to

$$j(t) = 4\pi DR^2 \left.\frac{\partial c}{\partial r}\right|_{r=R} = 4\pi DRc_0(1 + R/\sqrt{\pi Dt}) \qquad (1.15)$$

and becomes infinite at $t \to 0$.

In most cases this flux drops to final value extremely fast, so the full number of the reactant products predicted by the flux also has a finite value. But the very existence of this singularity at $t = 0$ "compromises" the theory, so the nature of this divergence must be thoroughly elucidated.

2. The boundary condition on the surface of the absorbing center

$$c(R,t) = 0 \qquad (1.16)$$

means that the center is an absolute sink (the probability of reaction P per one collision is equal to 1). However the physical sense of this condition is not quite clear: if the particle is not located on the surface of the sink, how can it get inside [25]? Besides, there are difficulties in formulating the boundary condition in the case when not all of the collisions between particles lead to a chemical reaction, i.e. at $P \neq 1$.

Collins and Kimball showed that both the postulates are interconnected, since the "divergence" of the flux at $t = 0$ can be considered the consequence of using boundary condition (1.16).

Without now going into a detailed discussion (see Chapter 4), and using as an example the sink problem, let us adduce several qualitative considerations which allow prediction of the character of a more general boundary condition [6,29]. It can be assumed that the probability of a reaction between mobile particle B and fixed sink A on the reacting sphere of radius R is proportional to the probability for particle B being located in the vicinity of the sphere in a layer from $r = R$ to $r = R + \Delta R$ (where ΔR is a small value). Then the flux on the sphere will be defined by the expression

$$j = \bar{k}c(R,t) \tag{1.17}$$

where $c(R,t)$ is the mean concentration of particles B in the given layer.

Using the general definition of the flux

$$j = 4\pi DR^2 \left(\frac{\partial c}{\partial r}\right)_{r = R} \tag{1.18}$$

we obtain the boundary condition

$$\bar{k}c(R,t) = 4\pi DR^2(\partial c/\partial r)_{r = R} \tag{1.19}$$

(index B is omitted).

The meaning of constant $\bar{k}$ can be established from Noyes' arguments [6]. Let us consider the case when flux j into the particle is constant. As the rate of the reaction is proportional to flux j, the following expression can be written for an arbitrary distance r

$$\frac{dc}{dr} = \frac{j}{4\pi Dr^2}$$

By integrating the expression we obtain

$$c(r) = c(r)\big|_{\infty} - j/(4\pi Dr) \tag{1.20}$$

where $c(r)|_{\infty}$ is the concentration of mobile particles at large distances from the absorbing sink.

From (1.17) and (1.18) one can establish that

$$c(R) = c(r)\big|_{\infty}/[1 + \bar{k}(4\pi DR)^{-1}] \tag{1.21}$$

Thus the concentration of particles on the reacting sphere is always lower than the mean concentration $C(t)$ of mobile particles far from the sink. If we consider that the number of sinks per volume unit is small, so that value $c(r)|_{\infty}$ characterizes mean concentration $C(t)$ of particles in the system at each time moment $t[c(r)|_{\infty} \to C(t)]$, then the experimentally measured rate constant k_{ex} of the reaction between particle A and center B is

$$k_{ex} = \frac{j}{C(t)} = \frac{j}{c(r)\Big|_{\infty}} = \frac{\bar{k}k_D}{k_D + \bar{k}}, \quad k_D = 4\pi DR \tag{1.22}$$

Within the limit $k_D \gg \bar{k}$ one obtains $k_{ex} = \bar{k}$, i.e., $\bar{k}$ defines the observed kinetic rate constant of the bimolecular reaction $\bar{k} \equiv k$. In the opposite case $k_D \ll \bar{k}$ one obtains $k_{ex} = k_D = 4\pi DR$, i.e. the rate constant calculated by Smoluchowski. Now the boundary condition assumes the anticipated form explicitly indicating the relation between diffusion and reaction.

Thus it follows that the basic concept of the new boundary condition is introducing the examination of two reactions (with constants k and k_D) as stages of a summary (total) reaction described by the equation

$$-dC_A(t)/dt = k_{ex}C_A(t)C_B(t) \tag{1.23}$$

Values $k_{ex}{}^{-1}$, K^{-1} and $k_D{}^{-1}$ can be treated as the corresponding kinetic resistances and (1.22) can be considered the law of addition of kinetic resistances

$$k_{ex}{}^{-1} = k^{-1} + k_D{}^{-1} \tag{1.24}$$

Generalizing the conclusion obtained for the Smoluchowski method, it can be asserted that the problem of boundary conditions is reduced to taking into account all the stages of the reaction in such a way that the diffusion equation (which generally speaking, characterizes one stage) would represent the whole process, permitting an analytic calculation of experimental rate constant k_{ex}.

This conclusion is most valuable in considering reversible reactions, since the direct application of the Smoluchowski method in this case leads to a seeming violation of the principle of detailed balance.

The reaction $C \rightarrow A + B$ reverse to (1.1) at first sight does not include reagent transport processes, and its observed rate constant $k_{ex}{}^{rev}$ should simply be equal to the corresponding kinetic rate constant k'. In this case equilibrium constant $K = k_{ex}(k_{ex}{}^{rev})^{-1}$ would depend on diffusion coefficient D [30]. However K, being the thermodynamic characteristic of the system, must not include parameters which determine the rate of establishing an equilibrium. Thus a purely diffusive control of the reaction leads to a seeming violation of the principle of detailed balance.

Naturally, if the effect of the transport process rate on the kinetics of reversible reactions is described consistently, no contradiction arises [31]. Such a description can be obtained by extending the boundary condition (1.19).

For convenience let us consider reversible reaction (1.1) assuming that the elementary act of a direct reaction results in annihilation of only an A type particle. Accordingly, the elementary act of a reserve reaction results in the release of an A type particle by a C particle. A particles are absorbed only upon collision with B particles, i.e., a direct reaction between two particles-reagents occurs only on the reacting surface when their centers are located at a minimal distance from each other. Analogously, A particles emerge only in immediate proximity to B particles.

Integrating both parts of Equation (1.2) throughout volume V (the volumes of the particles themselves are omitted) permits the expression

$$\int_V \frac{\partial c(\vec{r},t)}{\partial t} d\vec{r} = D \int_V \Delta c(\vec{r},t) \overrightarrow{dr} \tag{1.25}$$

Changing the order of integration and differentiation operations in the left part of Equation (1.25) lets us transform it in accordance with the Ostrogradski—Gausse theorem

$$\frac{\partial}{\partial t} \int_V c(\vec{r},t)\overrightarrow{dr} = D \int_S \nabla c(\vec{r},t)\overrightarrow{dS} \tag{1.26}$$

The left part of Equation (1.26) is equal to the change rate of the full number of A particles in the system, and the right part—to the diffusion flux through surface S, enveloping volume V. This surface is composed of all the reacting surfaces of B particles.

Since the rate of change of the number of particles and the diffusion flux are additive, it follows from Equation (1.26) that the mean value of particle A flux through the reacting surface surrounding particle B equals the mean reaction rate on the reacting surface.

The rate of a direct reaction on the reacting surface equals the product of rate kinetic constant k by particle A concentration on this surface. The rate of the reverse reaction is denoted by constant k'. Using the spheric symmetry of the problem, it is possible to frame a boundary condition using the average flux and mean particles concentration on the reacting surface

$$4\pi DR^2 \left.\frac{\partial c(r,t)}{\partial r}\right|_{r=R} = kc(R,t) - k' \tag{1.27}$$

In the equilibrium state the particle flux through the reacting surface is equal to zero, i.e., $kc(R,\infty) = k'$. Due to this, $K = k_{ex}/k_{ex}^{rev} = k_D/(k_D + k)c(R)$. If (1.21) is taken into account it becomes clear that this expression is independent of coefficient D.

The analysis of the problem of boundary conditions leads to the following conclusion. The taking into account of chemical reactions is in itself an independent problem. The simplest boundary conditions (1.16), (1.19) may not reflect [32,39] the specific character of all actual systems where the processes of diffusion transport play an important role in kinetics. The problem becomes considerably more complicated if one considers the structure of the interacting particles (effects of chemical heterogeneity and steric anisotropy), the character of the interaction (short-range and long-range effective potentials) of the colliding particles, and the specificity of particle interaction with the surface of condensed phases and regions of interphase boundaries (such specificity is assigned to the features of the structure and the dynamics of the surface layers of the condensed phase). The approach to framing boundary conditions for problems of diffusion kinetics that is developed below (see Chapter 4), in principle, allows taking into account the specificity of the interaction between both two colliding particles and a particle with the surface.

In conclusion it should be noted that the Smoluchowski problem with a type 1 boundary condition actually leads to a well-known problem in the theory of Markov processes, i.e., the problem of calculating the probability of reaching the boundary first by a diffusing particle [5,19,40]. This problem is of special interest in the event the reacting particles are connected by an interaction potential, while the diffusion coefficient depends on the time and location of the active particle in space. The solution of problems of reaching of a boundary first is, as a rule, connected with the application of the so-called conjugated equations [40–49]. These equations constitute the basis of one of the methods of DCR kinetics, which we shall call the method of conjugated equations.

1.4 THE METHOD OF CONJUGATED EQUATIONS

Let us introduce the basic functions characterizing Markov random processes. Let $\omega_1(x,t)$ be the absolute density probability that at time t the system is in point x, and let $\omega(x,t;\ x_0,t_0)$ be the conditional density of probability that at time t the system is in point x, given that at time t_0 the system was in point x_0 (thus function $\omega(x,t;\ x_0,t_0)$ can be considered the conditional density of probability of the corresponding transition.

The general differential equation for the transition probabilities which is known as the Einstein–Fokker–Kolmogorov–Planck equation, or simply the Fokker–Planck equation, is usually written in the form

$$\frac{\partial\omega(x,t;\ x_0,t_0)}{\partial t} = -\frac{\partial[A(x,t)\omega]}{\partial x} + \frac{1}{2}\frac{\partial^2[B(x,t)\omega]}{\partial x^2} \tag{1.28}$$

where $A(x,t)$ and $B(x,t)$ are functions called coefficients of drift and diffusion respectively.

The solution of Equation (1.28) should be non-negative, and normalized to 1, and should satisfy the initial condition

$$\omega(x,t_0;\ x_0,t_0) = \delta(x - x_0) \tag{1.29}$$

The physical sense of Equation (1.28) can be understood as follows. Let us assume that at time t_0 a large number (ensemble) of particles moving independently leaves point x_0. The flux of particles j is composed of a systematic ("hydrodynamic") flux $A\omega$ (where A is the rate of systematic movement in point x at time t) and a diffusion flux— $\partial B\omega/\partial x$, where $\frac{1}{2}B = D$ is the diffusion coefficient

$$j = A\omega - \frac{1}{2}\frac{\partial B\omega}{\partial x} \tag{1.30}$$

Then the Fokker–Planck equation is simply a continuity equation for particles of the ensemble

$$\partial \omega / \partial t + \operatorname{div} j = 0$$

which reflects the preservation of the number of particles. The determination of flux (1.30) plays the principle role in calculating the DCR rate constant if there is an interaction potential between particles.

The function of the transition probability $\omega(x,t;\ x_0 t_0)$, considered as a function of initial variables x_0, t_0 satisfies the conjugated equation

$$\frac{\partial \omega(x,t;\ x_0,t_0)}{\partial t_0} = -A(x_0,t_0)\frac{\partial \omega}{\partial x_0} - \frac{B(x_0,t_0)}{2}\frac{\partial^2 \omega}{\partial x_0^2} \tag{1.31}$$

This equation is sometimes called the reverse equation since it includes the initial time derivative $t_0 < t$ (the sign of the diffusion "coefficient" is negative in this equation).

Now let us establish the connection between functions $\omega_1(x,t)$ and $\omega(x,t;\ x_0,t_0)$. If at the initial time moment the initial distribution function $f(x_0,t_0)$ (instead of initial state x_0) is given, then it is possible to define the absolute probability of the transition $\omega_2(x,t;\ x_0,t_0)$:

$$\omega_2(x,t;\ x_0,t_0) = \omega(x,t;\ x_0,t_0) f(x_0,t_0) \tag{1.32}$$

If (1.32) is integrated with respect to x_0 the following expression is obtained

$$\omega_1(x,t) = \int \omega(x,t;\ x_0,t_0) f(x_0,t_0)\, dx_0 = \int \omega_2\, dx_0$$

Now it can be quite clearly seen that function $\omega_1(x,t)$ satisfies Equation (1.28) with the initial condition $\omega_1(x,t) = f(x,t_0)$, and the solution is non-negative and is normalized to 1.

It has already been mentioned that the problem of reaching of boundaries first [5] is of special interest for DCR kinetics. Let us formulate this in the following way. Let random variable $x(t)$ take at initial moment $t_0 = 0$ value x_0. We are interested in the statistic characteristics of the random time value of reaching first either the given boundary $x = a$ (or $x = b$), or any other boundary. We shall restrict ourselves by considering the simplest Markov process uniform in time, i.e., with function $\omega(x,t;\ x_0,t_0)$ depending solely on the difference $(t - t_0)$ and with time independent coefficients $A(x)$ and $B(x)$ in the Fokker–Planck equation. So we find function $\varphi(t,x_0)$, the probability of reaching the boundary points $x = a$ or $x = b$ during the time interval $(0,t)$ [46].

$$\varphi(t,x_0) = \text{Prob}\left\{ \begin{matrix} x = a \text{ or } x = b \\ \text{at least once} \\ \text{in time } t \end{matrix} \;\middle|\; \begin{matrix} \text{if at } t = 0 \\ \text{particle was} \\ \text{located at } x = x_0 \end{matrix} \right\}$$

The introduced function is an integral probability (so the density of the probability for reaching the boundaries at time t is $\partial\varphi/\partial t$). That is why the mean time of reaching one of the boundaries (a or b) is equal to

$$\bar{t}(x_0) = \int_0^\infty t \frac{\partial\varphi}{\partial t} dt \tag{1.33}$$

For $\varphi(t,x_0)$ the equation has the form

$$\frac{\partial\varphi(t,x_0)}{\partial t} = A(x_0)\frac{\partial\varphi}{\partial x_0} + B(x_0)\frac{\partial^2\varphi}{\partial x_0^2} \tag{1.34}$$

This equation is also called a conjugated one. Similar to Equation (1.31), it has the same parameters of $A(x_0)$ and $B(x_0)$ but the left part contains the derivatives on time t, not initial time t_0.

The initial and boundary conditions for Equation (1.34) are determined by the physical meaning of value $\varphi(t,x_0)$. If at the initial time moment the particle is not on the boundary, then

$$\varphi(0,x_0) = 0, \qquad a < x_0 < b \tag{1.35}$$

But if at $t = 0$ one of the boundaries is occupied ($x = a$ or $x = b$) for each $t > 0$ we have

$$\varphi(t,a) = \varphi(t,b) = 1, \qquad t > 0 \tag{1.36}$$

The use of conjugated equations permits simplifying considerably the solution of problems of DCR theory in a number of cases. Quite often it is sufficient to obtain nondetailed information about a system (held in the distribution function), as well as to find the average characteristics of processes (for example, the probabilities of particle survival that is connected with the value of radiation escape or the mean lifetime of active particles determined by Equation (1.33), etc.).

Let us exemplify the direct application of conjugated Equation (1.34) in studying DCR kinetics, assuming for the sake of simplicity that $D = B/2$ (independent of x and t), and $A = 0$.

1. First we shall consider a simple problem of the recombination of the particle pair, "disappearing" upon first encounter [5]. If the problem has spheric symmetry, Equation (1.34) assumes the form

$$\partial(r_0\varphi)/\partial t = D\partial^2(r_0\varphi)/\partial r_0^2 \tag{1.37}$$

and it can be solved with initial condition $\varphi(0,r_0) = 0$ and boundary condition (1.36)

$$\varphi(t,R) = 1 \tag{1.38}$$

Now let us choose the second boundary condition in the form

$$\varphi(t,\infty) = 0 \tag{1.39}$$

which shows the impossibility of reaching an infinitely far boundary in finite time.

The solution of problem (1.37)–(1.39) is well known [5,49]:

$$\varphi(t,r_0) = \frac{R}{r_0}\left[1 - \Phi\left(\frac{r_0 - R}{\sqrt{4Dt}}\right)\right] \tag{1.40}$$

where $\Phi(z)$ is the probability integral.

The conditional probability of particle recombination per time unit is prescribed by value $\partial\varphi/\partial t$

$$\frac{\partial\varphi}{\partial t} = \frac{R(r_0 - R)}{r_0\sqrt{\pi D t^3}} \exp\left[-\frac{(r_0 - R)^2}{4Dt}\right] \tag{1.41}$$

This expression can be used to determine the absolute probability of recombination in the event of averaging over initial particle position. In the case of a random initial arrangement of particles, averaging should bring us to expression (1.6) of the Smoluchowski method. However, the "divergence" of flux $j(t)$ at $t \to 0$ in the Smoluchowski method demands a more detailed discussion of the averaging procedure. At all values of t (at $t = 0$ included) expression (1.40) and (1.41) hold for the open interval $r_0 > R$, where $\partial\varphi/\partial t \to 0$ at $t \to 0$. Averaging over all initial particle positions is given by integral

$$\overline{\frac{\partial\varphi}{\partial t}} = V^{-1} \int_{R+\varepsilon}^{\infty} \frac{\partial\varphi}{\partial t}\, 4\pi r_0^2\, dr_0$$

throughout the volume of the system excluding ε-neighborhood of point $r_0 = R$. Since at $r_0 \to \infty$, $\partial\varphi/\partial r_0 \to 0$, the resulting equation is written

$$\overline{\frac{\partial\varphi}{\partial t}} = \frac{4\pi RD}{V}\left[1 - \Phi\left(\frac{\varepsilon}{\sqrt{4Dt}}\right) + \frac{R\exp(-\varepsilon^2/4Dt)}{\sqrt{\pi D t}}\right] \tag{1.42}$$

Of special interest is one specific feature of expression (1.42). Although it is accepted that the "canonical" method of Smoluchowski is compromised by the existence of flux divergence into the particles at short times [see (1.15)], the method of the conjugated equations leads to a reasonable solution $j \to 0$ at $t \to 0$. For time $t \gg \varepsilon^2 D^{-1}$ the well-known expression (1.6) is obtained

$$\overline{\frac{\partial\varphi}{\partial t}} = j = 4\pi DR(1 + R/\sqrt{\pi D t})V^{-1}$$

Evidently the exclusion of ε-neighborhood when calculating the flux constitutes the difference between (1.6) and (1.42).

The initial equations are the same in both the Smoluchowski method and the method of conjugated equations, but the boundary conditions are different. Smoluchowski's condition $c(R) = 0$ is inconsistent: how can particles be absorbed if they are "not" on the boundary? Condition (1.38) in the method of the conjugated equations is adequate, but the conjugated equation is considered strictly in the interval excluding boundaries $a < r_0 < b$. It was by consistently taking this factor (introduction of $\varepsilon \to 0$) into account that made it possible to obtain the correct answer.

This discussion leads to the following conclusion. The deficiencies of the Smoluchowski method can be avoided if the equation is adequately analyzed in ε-neighborhood. This indicates the necessity of considering distances where diffusion approximation does not "work", and of analyzing the processes of discrete random walks. This will be done in Chapter 4 in discussing the general form of boundary conditions.

2. Let us study the time dependence [49] of the number of particles $n(t)$ in the system of recombinating particles after averaging over initial particle distribution. The corresponding averaging involves great mathematical difficulties. However, it turned out that it is possible to formulate the problem in terms of averaged pair probabilities of collisions between recombinating particles.

Using function $\varphi(t,r)$, the probability for reaching the boundary during time t, let us introduce average probability $\overline{W}(t)$ of a collision in time t of an arbitrary particle pair

$$\overline{W}(t) = \int_{\alpha}^{\infty} \varphi(t,r)\rho(r)dr \tag{1.43}$$

where $\rho(r)$ is the density of probability of relative distances between particles at $t = 0$. Thus we obtain an equation connecting $n(t)$ with averaged probability $\overline{W}$. First let us consider the discrete model of particle disappearance, assuming that at discrete moments $\tau_i (i = 1, 2, \ldots)$ the number of particles decreases by value Δn_i. Obviously after averaging over the initial distribution the probability for an arbitrary particle to escape a collision in the environment of n particles is $[1 - \overline{W}(t)]^n$. Taking into account the fact that the number of particles in the system decreases, this probability can be presented in the form

$$\omega(t) = [1 - \overline{W}(t)]^{n(t)} \prod_i [1 - \overline{W}(\tau_i)]^{-\Delta n_i} \tag{1.44}$$

Multiplication of both parts in (1.44) by n_0—the full number of particles at $t = 0$ yields an equation in a discrete model

$$n(t) = n_0[1 - \overline{W}(t)]^{n(t)} \prod_i [1 - \overline{W}(\tau_i)]^{-\Delta n_i}$$

The continuous analogue of this equation has the form

$$n(t) = n_0[1 - \overline{W}(t)]^{n(t)} \exp\left\{-\int_0^t d\tau \frac{dn(\tau)}{d\tau} \ln[1 - \overline{W}(\tau)]\right\} \tag{1.45}$$

which leads to the solution

$$n(t) = n_0\{1 - n_0 \ln[1 - \overline{W}(t)]\}^{-1} \tag{1.46}$$

It should be noted that $\overline{W} \ll 1$ since probability averaging $\varphi(t,r)$ is carried out over a large volume of the system. That is why (1.46) can be rewritten in the form

$$n(t) = n_0[1 + n_0\overline{W}(t)]^{-1}.$$

The result obtained is a general solution which can be used to describe the disappearance process of particles with an arbitrary initial distribution. Let us apply it to an analysis of particle recombination on the surface which is quite difficult using the Smoluchowski method [50]. We assume that two particles upon collision become nonactive and leave the ensemble of reacting particles.

In describing processes occurring on the surface, the following equation can be derived (in case of cylindric symmetry of the problem) for probability $\varphi(t,r)$

$$\frac{\partial\varphi(t,r)}{\partial t} = \frac{D}{r}\frac{\partial}{\partial r}\left(r\frac{\partial\varphi}{\partial r}\right) \tag{1.47}$$

with boundary conditions

$$\varphi(t,a) = 1, \qquad \varphi(t,\infty) = 0 \tag{1.48}$$

and initial condition

$$\varphi(0,r) = 0, \qquad r > a \tag{1.49}$$

The solution of equation (1.47) satisfying conditions (1.48) and (1.49) has the form

$$\varphi(t,r) = 1 - \frac{2}{\pi}\int_0^\infty dx \exp\left(-\frac{Dt}{a^2}x^2\right)\frac{[I_0(x)N_0(rx/a) - N_0(x)I_0(rx/a)]}{x[I_0^{\,2}(x) + N_0^{\,2}(x)]}$$

where $I_0(x)$ and $N_0(x)$ are Bessel function and Neumann zero order function respectively.

At random initial particle distribution on the surface of area S, probability $\varphi(t,r)$ is averaged in (1.43) with function $\rho(r) = 2\pi r S^{-1}$. As a result the following equation can be derived for the number of particles at time t

$$\frac{n_0}{n(t)} = 1 + \frac{8\rho_0 a^2}{\pi} \int_0^\infty \frac{\{1 - \exp[-(Dt/a^2)x^2]\}}{x^3[I_0^2(x) + N_0^2(x)]}\, dx \tag{1.50}$$

where ρ_0 is the initial surface density of particles.

Thus, using the method of the conjugated equations makes it possible to avoid the difficulties inherent in the Smoluchowski method.

Most often conjugated equations are applied for finding either the mean time $\tau(\vec{r}_0)$ of the reacting boundary from the initial state $\vec{r}_0$ or the mean time τ if the particle stays in some region, for example in the reaction zone. Knowing how to calculate these values considerably simplifies designing more efficient approximation methods for solving various problems of diffusion controlled kinetics.

1.5 DIFFUSION CONTROLLED REACTION OF ANISOTROPIC MOLECULES [37,51–55]

The elementary act of chemical transformation as a rule occurs at a definite particle orientation at the moment of their "contact". In this case one can speak about chemical anisotropy of reagents which is characterized by effective steric factor $f_{eff} \leqslant 1$.

For describing DCR of anisotropic molecules, the model of a black spot [56] is used. The model [56] assumes that the reacting region covers only a certain part ("a black spot") of the surface of the particle. The particle collision brings about a chemical transformation only if these black spots touch each other.

Using the Smoluchowski method for finding rate constant k_a of such reactions requires solving diffusion equations in multidimensional space which includes angular variables of particles with the corresponding boundary conditions with anisotropic reactivity on a surface. Finding such solutions is rather a complicated problem which necessitated working out physical approximations for simplifying the problem.

The most interesting approximation is connected with the following assumption: the reaction zone is a "reaction" layer of volume V such that the mean time the system (or the point presenting it) stays in it is τ. Then rate constant k_a of the process is determined by the expression

$$k_a = V/\tau \tag{1.51}$$

In the simplest case of chemically isotropic reagents, the reaction zone is a thin spheric layer of width $\Delta \ll R$ (where R is the radius of the reaction surface), so $V = 4\pi R^2\Delta$, $\tau = R\Delta/D$ and definition (1.51) bring to the usual expression $k_D = 4\pi DR$.

Let us consider the case of interaction between isotropic and "anisotropic" molecules with a reacting layer denoted by value Δ and by angular dimensions of spot θ_m, so that the coordinates of the radius—vector of the point in the reaction zone satisfy the conditions

$$R < r < R + \Delta, \qquad 0 < \theta < \theta_m$$

The effective steric factor f_{eff} in this case depends only on θ_m and is

$$f_{eff} = \sin^2(\theta_m/2)$$

It is equal to the ratio of the reaction spot area of the anisotropic molecule to the full surface of the second molecule. The expression for the volume of reaction zone V then assumes the form

$$V = 4\pi R^2 \Delta f_{eff} \tag{1.52}$$

and the problem of calculating rate constant k_a is reduced to calculating τ—the mean time the system stays in the reaction zone. In the general case [57] it can be expressed with the aid of conditional probability $\omega(\vec{r}, t; \vec{r}_0 t_0)$, obeying the equation of type (1.28):

$$\tau = \frac{1}{V} \int_V \int_V d\vec{r}\, d\vec{r}_0 \int_0^\infty \omega(\vec{r}, t; \vec{r}_0, D)\, dt \tag{1.53}$$

Function $\omega(\vec{r}, t; \vec{r}_0, 0)$ must satisfy initial condition (1.29) and also the boundary condition

$$\partial\omega/\partial r\Big|_{r=R} = 0$$

The latter reflects the impossibility of a particle being located inside the spheric volume with $r < R$. We shall only consider the simplest case of Equation (1.28) when $B/2 = D$, $A = 0$ and τ can be calculated using Green's function $G(\vec{r}, \vec{r}_0)$ of diffusion movement

$$\tau = \frac{1}{V} \int_V \int_V d\vec{r}\, d\vec{r}_0 G(\vec{r}, \vec{r}_0) \tag{1.54}$$

(integrating is carried out over volume V of the reaction zone).

Function $G(\vec{r}, \vec{r}_0)$ has been analyzed in [57,58] for the model of a thin spheric layer, when it can be assumed that $r = r_0 = R$, it has the form

$$G(\theta, \varphi; \theta_0, \varphi_0) = \frac{1}{4\pi D R} \left\{ \left(\frac{2}{1 - \cos\gamma} \right)^{1/2} - \ln\left[1 + \left(\frac{2}{1 - \cos\gamma} \right)^{1/2} \right] \right\} \tag{1.55}$$

where θ, φ are angular variables of a spheric coordinate system; γ is the angle between vectors $\vec{r}$ and $\vec{r}_0$.

The qualitative effects of chemical anisotropy on rate constant k_a can be easily illustrated in the case of strong anisotropy when $f_{eff} \ll 1$, i.e. $\theta, \theta' \leqslant \theta_m \ll \pi$ (for more details see [51–56]).

At $f_{eff} \ll 1$ the logarithmic term in Eq. (1.55) can be ignored, if it is assumed that

$$\cos\gamma = 1 - \frac{\theta^2 + \theta_0^2}{2} + \theta\theta_0 \cos(\varphi - \varphi_0)$$

and it becomes possible to calculate the mean time $\tau(R, \theta_0, {*}_0)$ the system stays in the reaction zone if at initial time it was in point $\vec{r}_0 = (R, \theta_0, \varphi_0)$

$$\tau(R, \theta_0, \varphi_0) = \frac{1}{4\pi DR} \int_R^{R+\Delta} \int_0^{\theta_m} \int_0^{2\pi} \times$$

$$\times \frac{r^2 \sin\theta dr d\theta d\varphi}{\left[(\theta + \theta_0)^2 - 4\theta\theta_0 \cos^2\left(\frac{\varphi}{2} - \frac{\varphi 0}{2}\right)\right]^{1/2}} = \frac{4R\Delta}{\pi D} f^{1/2}_{\text{eff}} E(\theta_0/\theta_m)$$

where $E(z) = \int_0^{\pi/2} (1 - z^2 \sin^2\alpha)^{1/2} d\alpha$ is the full elliptic integral of second order.

Result (1.55) is independent of φ_0 because of the axial symmetry of the problem, and averaging over θ_0 yields expression

$$\tau = \frac{16}{3\pi} \frac{R\Delta}{D} \sqrt{f_{\text{eff}}} \tag{1.56}$$

An important relationship is obtained upon substituting (1.52) and (1.56) into (1.51)

$$k_a = 4\pi DR_{\text{eff}} \tag{1.57}$$

$$R_{\text{eff}} = R \frac{3}{16} \pi\sqrt{f_{\text{eff}}}$$

The rate constant appears to be proportional to the value $\sqrt{f_{\text{eff}}}$ greater than the relative area f_{eff} of the spot. This fact can be interpreted as a result of reaction anisotropy averaging due to relative movement of molecules.

Expression (1.57) is of a general character [53]; the approximations of the model are quite fully discussed in [53,55,59,60]. It should be noted, however, that the model considered "loses" an essential feature of the Smoluchowski model: the possibility of determining the dependence of the DCR rate constant on time.

The renormalization of the "interaction radius" value in the expression for diffusion rate constant (1.57) can be interpreted as a consequence of the change of the effective interaction between particles resulting from the chemical anisotropy of molecules. And indeed, a DCR analysis of charged particles [61,62] also presents the effective rate constant as a diffusion constant $k_{ex} = 4\pi DR_{\text{eff}}$, where R_{eff} depends on the interaction potential. Renormalization of effective rate constants can also be caused by other (besides reactivity anisotropy and the long-range particle potential) factors, in particular, the multistaged character of chemical processes.

1.6 THE EFFECT OF TRANSFER PROCESSES ON THE KINETICS OF MULTISTAGED CHEMICAL TRANSFORMATIONS

Usually diffusion kinetics considers the simplest mono- and bimolecular processes, whereas multistaged chemical transformations are considered within the framework of formal kinetics, completely ignoring processes of reagent transport. However, in the case of low particle mobility (at least at one stage) it become absolutely necessary to take into account the diffusive approaching of reagents (at initial stages of the process it is always necessary).

Taking into account the diffusion stages of approaching particles in considering multistage chemical processes leads to renormalization of the corresponding chemical constants. In the nonstationary case it determines the characteristic time dependence. Let us analyze this situation in the general case of a multistage chemical transformation [63], in which the transfer effects are relevant only at one of the stages—reaction of C particles on active centers N (assuming that equations of formal kinetics hold for all the other stages).

Let us assume that N centers form a regular three-dimensional lattice with period $2L$. Let C particles appear upon the decomposition of A molecules

$$A \xrightarrow[k_1']{k_1} C + P_{(I)}$$

and participate in other reactions, for example

$$C + B \underset{k_2'}{\overset{k_2}{\rightleftarrows}} P_{(2)} + P_{(3)}$$

Then after breaking up the whole of the volume into cells, each containing one center of N type, and approximating these regions by spheres of radius L (cell method, see Chapter 3) the concentration $\rho(r, t)$ of C particles within one cell is determined by the equation

$$\frac{\partial \rho}{\partial t} = \frac{D}{r} \frac{\partial^2 (r\rho)}{\partial r^2} + k_1 \bar{a} - k_2 \bar{b} \rho - k_1' \bar{\rho}_1 \rho + k_2' \bar{\rho}_2 \bar{\rho}_3 \tag{1.58}$$

at the following boundary conditions

$$D \left. \frac{\partial \rho}{\partial r} \right|_{r = R} = \chi \rho(R, t); \quad \left. \frac{\partial \rho}{\partial r} \right|_{r = L} = 0 \tag{1.59}$$

where $\bar{a}(t)$, $\bar{b}(t)$, $\bar{\rho}_k(t) (k = 1,2,3)$ are respectively the average concentrations of particles A, B, P_k; R—reaction radius; χ—rate constant of the chemical reaction.

We shall consider the kinetics of the process in a particular case, when $\rho(r, 0) = \rho_0$.

After introducing mean concentration $\bar{\rho}(t)$ of C particles

$$\bar{\rho}(t) = \frac{3}{L^3} \int_R^L \rho(r,t) r^2 dr$$

it then follows from (1.58), (1.59)

$$\frac{d\bar{\rho}}{dt} = -\ 4\pi R^2 \chi n \rho(R,t) + k_1 \bar{a} - k_2 \overline{b\rho} - k_1{}' \bar{\rho}_1 \bar{\rho} + k_2{}' \bar{\rho}_2 \bar{\rho}_3 \tag{1.60}$$

$$n = \left(\frac{4}{3}\pi L^3\right)^{-1}$$

The corresponding equation of formal kinetics has the form

$$\frac{d\bar{\rho}}{dt} = -k\bar{\rho}(t) n + k_1 \bar{a} - k_2 \overline{b\rho} - k_1{}' \bar{\rho}_1 \bar{\rho} + k_2{}' \bar{\rho}_2 \bar{\rho}_3 \tag{1.61}$$

$$k = 4\pi \chi R^2$$

So it follows from (1.60) and (1.61) that taking into account the diffusion of C particles has resulted in the renormalization of rate constant k in introducing effective constant

$$k_{\text{eff}}(t) = k\rho(r,t)/\bar{\rho}(t)$$

Essentially, mathematical modelling of the multistage physical-chemical transformations should be reduced to a selfconsistent determination of dependencies $k_{\text{eff}}(t)$ and mean concentrations of reacting particles $\bar{a}(t)$, $\bar{b}(t)$,... from equations of formal kinetics and Equation (1.60). Obviously, computers must be used in order to complete such a program.

Let us consider only the simple case when the concentration of A and B particles does not practically change in the course of a chemical reaction of C particles on N centers (we assume that $k_1{}' = 0$, $k_2{}' = 0$ and $\bar{a} = a_0$, $\bar{b} = b_0$). Then solving (1.58) under these conditions and using (1.59) for $k_{\text{eff}}(t)$ (at $L \gg \{R; \sqrt{D/k_2 b_0}\}$):

$$k_{\text{eff}}(t) = k\left[1 + s\exp\left(-\frac{\gamma^2}{R^2}Dt\right)\right]^{-1}\left\{\frac{1+\gamma}{\alpha+\gamma} + \frac{s}{\alpha}\exp\left(-\frac{\gamma^2}{R^2}Dt\right) + \right.$$

$$+ \frac{\alpha - 1}{\alpha}\left(s + \frac{\alpha^2}{\alpha^2 - \gamma^2}\right)\exp\left[(\alpha^2 - \gamma^2)\frac{Dt}{R^2}\right]\operatorname{erfc}\left(\frac{\alpha\sqrt{Dt}}{R}\right) -$$

$$\left. - \frac{(\alpha - 1)\gamma}{\alpha^2 - \gamma^2}\operatorname{erfc}\left(\frac{\gamma}{R}\sqrt{Dt}\right)\right\}$$

where

$$\gamma = R\sqrt{k_2 b_0 / D}, \qquad \alpha = 1 + \frac{\chi R}{D}, \qquad s = \rho_0 \frac{k_2 b_0}{k_1 a_0} - 1$$

The stationary case corresponds to the limit $t \to \infty$ and rate constant is

$$k^{st}_{eff} = k(1 + \gamma)/(\alpha + \gamma)$$

For calculating concentrations a_0 and b_0, which determine k_{eff}, it is necessary to solve a system of differential equations for chemical kinetics conforming to the considered multistage process replacing k by k_{eff}. It is also necessary to solve the transcendental equations obtained for a_0 and b_0. Then all mean concentrations of reacting particles as well as the rates of processes in which C particles are not directly engaged may turn out to be dependent on parameters D and R of the diffusion stage. It should be noted that transfer effects for other stages can be taken similarily into account only if the corresponding differential equations are not "hooked", i.e., they can be reduced to equations of type (1.61) for mean concentrations.

All this should be taken into consideration in an analysis of the kinetic data at low reagent mobility. It should be brought out that in this analysis the effective constants become dependent on the rate of nonstationarity of the process. This fact is usually noted for reactions occurring in sharply non-stationary conditions. However, the quantitative relations of such effective constants with true parameters of the process (rate of chemical reaction, diffusion parameters) are not as a rule considered. It should be said in conclusion that the approximation used (the breaking of the volume into cells with boundary condition $\partial\rho/\partial r|_{r=L} = 0$) cannot be used when the non-stationary diffusion kinetics of surface processes is being considere This is connected with the specificity (mentioned in Section 1.4) of two-dimensional diffusion—its weak (logarithmic)—dependence of probability $W(r,t)$ of a diffusion collision between two particles located at distance r at initial time $t = 0$

$$W(r,t) \simeq 1 - \ln \frac{r}{R} \Big/ \ln (Dt/R^2)^{1/2} \qquad \text{if} \qquad t \gg R^2/D \tag{1.62}$$

It follows from (1.62) that a diffusing particle reaches neighboring and distant centers with practically the same probability (unlike the three-dimensional case when the reaction mostly occurs on the nearest center). This explains why the study of non-stationary kinetics in the two-dimensional case requires solving the problem in the entire space.

The example given points to a "non-standard" factor causing the renormalization of the effective reaction rate constant. It has already been mentioned that such a renormalization can be caused by other factors such as the "organization" of the chemical reaction elementary act.

1.7 THREE-MOLECULAR REACTIONS [64]

An analysis of the simplest models of three-molecular reactions

$$A + B + C \longrightarrow \text{product} \tag{1.63}$$

also leads to the renormalization of standard expressions for DCR rate constants. Let us assume that concentrations $\rho_A(r,t)$ and $\rho_B(r,t)$ of particles A and B at distance r from the center of particle C* at time t satisfy the diffusion equation

$$\frac{\partial \rho_i(r,t)}{\partial t} = D_i \Delta \rho_i(r,t) \qquad (i = A,B) \tag{1.64}$$

Suppose that the elementary act of a chemical reaction occurs at a triple collision when the distance between the centers of A and C particles is R_1, whereas the distance between the centers of B and C particles is R_2. Values R_1 and R_2 are the sums of radii R_i of the corresponding particles $(i = A,B,C)$, i.e. $R_1 = R_A + R_C$, $R_2 = R_C + R_B$. We shall only consider the case of low concentrations of reagents $\rho_i(\infty,t) \equiv n_i(t)$ when $n_i R_i^3 \ll 1$. When deducing an equation for the reaction rate with respect to component $i = A,B$, the results obtained for bimolecular diffusion controlled reactions can be applied. In the absence of component B, the rate of mean concentration decrease of component A would be proportional to flux Φ_A of particles A into sphere R_1

$$\frac{dn_A}{dt} = 4\pi R_1^2 D_1 n_C \left.\frac{\partial \rho_A}{\partial r}\right|_{R = R_1} = -n_C \Phi_A$$

An analogous expression could also be written for component B in the absence of A.

Since particles A and B simultaneously disappear on C particles (the origin of coordinates is placed in the center of a C type particle), it follows that

$$\Phi_A = \Phi_B = \Phi \tag{1.65}$$

$$\frac{dn_i}{dt} = - n_C \Phi \qquad (i = A,B) \tag{1.66}$$

Equations (1.65) and (1.66) do not concretize the trimolecular character of the reaction (separate reactions $A + C$ and $B + C$ are possible). That is why an additional assumption should be introduced

---

*The role of B can be performed, for example, by a "particle" of free volume of medium.

$$\Phi = k\rho_1(R_1,t)\rho_2(R_2,t) \tag{1.67}$$

i.e., the probability of a reaction on C particle is proportional to the density product of particles A and B on the corresponding reaction surface. In (1.67) value k is "the true" reaction rate constant.

In searching for an explicit expression of flux using (1.64), (1.65) and (1.67) we shall reduce the consideration to a quasistationary regime at $t \gg R_i^2/D$ when

$$\rho_A(r) = \frac{A_1}{r} + B_1, \qquad \rho_B(r) = \frac{A_2}{r} + B_2 \tag{1.68}$$

Constants B_1 and B_2 are chosen so that $\rho_i \to n_i$ $(i = A, B)$ at $r \to \infty$. Then to determine constants A_1 and A_2, it is sufficient to apply (1.65) and (1.67), (1.68). The result is

$$\Phi = \frac{2\pi}{k}\Big\{4\pi D_1D_2R_1R_2 + k(n_1R_1D_1 + n_2R_2D_2) - [(4\pi D_1D_2R_1R_2 + kn_1R_1D_1 + kn_2R_2D_2)^2 - 4k^2D_1D_2R_1R_2n_1n_2]^{1/2}\Big\} \tag{1.69}$$

Unlike the corresponding value for bimolecular diffusion controlled reactions, the quasistationary value of flux Φ is nonlinear with respect to the mean concentration of reagents $n_1 = n_A$ and $n_2 = n_B$, i.e., taking into account the final rate of reagent diffusion in the case of trimolecular reactions leading to a deviation from the mass action law. It also follows from (1.69) that in the case of trimolecular reactions, the criteria of diffusion (or kinetic) control contains current reagent concentrations. Generally speaking, one and the same trimolecular reaction can pass over time from a diffusive to a kinetic regime.

Let us analyze the consequences following from (1.69). The first to be considered will be the case of small concentration values of reagent (B)

$$kn_2 \equiv kn_B \ll 4\pi D_1R_1 \tag{1.70}$$

Expanding the right part of expression (1.69) in terms of the small parameter $kn_2/4\pi D_1R_1$ and using only the first two members of the expansion yield

$$\Phi = 4\pi D_2R_2kn_1n_2\,/(4\pi R_2D_2 + kn_1) \tag{1.71}$$

Physically speaking, (1.71) means that if condition (1.70) is met the distribution of reagent A may be considered uniform (its transporting is "instantaneous"). Then the kinetics of reaction (1.63) may be described as the kinetics of bimolecular reactions with "real" rate constant kn_1.

If the concentration of reagent A is also sufficiently low, i.e., $kn_1 \ll 4\pi R_2D_2$, $\Phi = kn_1n_2$, the reaction rate obeys the laws of formal kinetics

$$\frac{dn_i}{dt} = -n_C\Phi = -kn_An_Bn_C$$

Whereas if the concentration of reagent A is sufficiently high, i.e.,

$$kn_1 \gg 4\pi D_2R_2$$

(the reaction proceeding in the diffusion regime with respect to reagent A and in the kinetic regime with respect to reagent B) flux Φ is determined solely by the transport rate of B particles

$$\Phi = 4\pi R_2D_2n_2$$

Accordingly, the reaction kinetics obey the laws of second order

$$\frac{dn_i}{dt} = -4\pi D_2R_2n_Bn_C \qquad (i = A,B,C)$$

The answer is less obvious in the case of large concentration values of reagents A and B, when

$$kn_1 \gg 4\pi D_2R_2$$
$$kn_2 \gg 4\pi D_1R_1 \tag{1.72}$$

(trimolecular reactions proceeding in the diffusion regime with respect to reagents A and B). Ignoring terms $4\pi R_1R_2D_1D_2$ (in comparison to terms $kR_iD_in_i$) in the expression (1.69) for Φ we have

$$\Phi = 4\pi \min\{R_iD_in_i\}$$

Generally speaking the obtained solution holds if particles of reagent C can be considered stationary. The character of the solution obtained points to the fact that in the general case when $D_C \neq 0$ the reagent C satisfying the inequality

$$n_CR_CD_C < R_iD_in_i \qquad (i = A,B)$$

should be chosen as an "immovable" (fixed) one.

Thus, if conditions (1.72) are met the reaction rate is determined by second order laws

$$\frac{dn_i}{dt} = -4\pi \min\{R_iD_in_i\}$$

In the case of trimolecular reactions of the type

$$3A \longrightarrow \text{Product}$$

the expression for flux (1.69) can be written in the form

$$\Phi = \frac{4\pi DR}{k}\left[2\pi DR + kn - \sqrt{(2\pi DR)^2 + 4\pi DkRn}\right] \tag{1.73}$$

If at low values of reagent density

$$kn \ll 2\pi DR$$

the root in (1.73) is expanded in parameter $kn/2\pi DR$ and only the first three members of the expansion are retained, then

$$\frac{dn}{dt} = -\ kn^3$$

i.e., the kinetics are determined by the third order law of formal kinetics. At high reagent density (the diffusion regime) of the reaction

$$kn \gg 2\pi DR$$

the kinetics are determined by the law of second order

$$\Phi = 4\pi RDn$$

$$\frac{dn}{dt} = -\ 4\pi RDn^2$$

1.8 SUBSTANTIATION OF DIFFUSION APPROXIMATION

1.8.1 The Langevin Equation

Methods of describing random particle movement due to thermal motion in the system were first developed by Smoluchowski and Einstein in reference to the problem of Brownian motion. The initial prerequisite [19] of such an analysis is that increment $\Delta\vec{u}$ in the velocity of a macroscopic particle in time Δt can be presented as a sum of two components if Δt exceeds by at least several times the period of elementary fluctuations. One of the contributions—$\vec{F}\Delta t$, corresponds to the action of the external force field, the other characterizes fluctuating quantity $\delta\vec{u}(\Delta t)$

$$\Delta\vec{u} = \vec{F}\,\Delta t + \delta\vec{u}(\Delta t) \tag{1.74}$$

For the corresponding increment $\Delta\vec{r}$

$$\Delta\vec{r} = \vec{u}\,\Delta t \tag{1.75}$$

where $\vec{u}$ is the momentary particle velocity.

Two parts should be singled out in component $\delta\vec{u}(\Delta t)$: $-\beta_0\vec{u}\Delta t$ corresponding to the acceleration produced by hydrodynamic friction* and the fluctuating part $\vec{B}(\Delta t)$—the acceleration a particle can experience within the given period Δt under the effect of a rather large number of collisions with separate molecules of the liquid environment

$$\delta\vec{u}(\Delta t) = -\beta_0\vec{u}\Delta t + \vec{B}(\Delta t) \qquad (1.76)$$

$$\vec{B}(\Delta t) = \int_t^{t+\Delta t} \vec{A}(\xi)d\xi \qquad \Delta t \ll t \qquad (1.77)$$

where $\vec{A}(t)$ is the fluctuating force due to the action of medium molecules on the Brownian particle.

This force is independent of $\vec{u}$ and changes considerably faster than $\vec{u}$. Thus deriving an equation for particle movement leads to the Langevin equation

$$d\vec{u}/dt = -(\beta_0\vec{u} - \vec{F}) + A(t) \qquad (1.78)$$

Taking into account the above-mentioned restrictions on function $\vec{A}(t)$, an analysis of stochastic Equation (1.78) shows that the solution should be understood in the sense of probability theory. For example, it may be required to determine distribution $W(\vec{u},t;\ \vec{u}_0)$ which assigns the probability of finding velocity $\vec{u}$ at time t, if at $t = 0$, the Brownian particle has velocity $\vec{u}_0$.

The formal solution of Equation (1.78) has the form

$$\vec{u} - \vec{u}_0 e^{-\beta_0 t} = e^{-\beta_0 t}\int_0^t e^{\beta_0\xi}\vec{A}(\xi)d\xi \qquad (1.79)$$

So the statistical properties being sought of expression

$$\vec{u} - \vec{u}_0 \exp(-\beta_0 t) \qquad (1.80)$$

are determined by the corresponding properties of the right part (1.79)

$$e^{-\beta_0 t}\int_0^t e^{\beta_0\xi}\vec{A}(\xi)d\xi = \sum_j e^{\beta_0 j(\Delta t - t)}\vec{B}(\Delta t) \qquad (1.81)$$

---

*When analyzing the transport of spheric particles of radius a and mass m, parameter β_0 is determined by Stokes law: $\beta_0 = 6\pi a\eta/m$, where η is the viscosity coefficient of the liquid environment.

To find the probability properties of difference (1.80) the probability density of $\vec{B}(\Delta t)$ should be assigned.

Obviously, $\vec{B}(\Delta t)$ can be presented as a sum of random vector quantities $\vec{b}_i$, which characterize the "elementary" accelerations of a Brownian particle originating from its separate encounters with molecules of the medium

$$\vec{B}(\Delta t) \equiv \vec{B}_N = \sum_{i=1}^{N} \vec{b}_i \qquad N = \Delta t/\tau$$

where τ is the time between encounters.

Let us assume that random quantities $\vec{b}_i$ are characterized by some distribution function with a finite dispersion. In this case at a sufficiently large N, distribution function $\omega[\vec{B}(\Delta t)]$ should tend to be that given by a normal law, i.e., assume a Gaussian form according to the central limit theorem of the probability theory [65]

$$\omega[\vec{B}(\Delta t)] = (4q\pi\tau N)^{-3/2} \exp\left(-\frac{|\vec{B}(\Delta t)|^2}{4q\Delta t}\right) \tag{1.82}$$

where q is a parameter.

In the case of Brownian macroparticle motion in liquid, the velocity distribution of these particles at $t \to \infty$ must be Maxwellian. It can be easily shown [19] that this is achieved (1.82) if $q = \beta_0 kT/m$. Choosing $\omega[\vec{B}(\Delta t)]$ in form (1.82) one can prove that

$$W(\vec{u}, t; \vec{u}_0) = \left[\frac{m}{2\pi kT(1 - e^{-2\beta_0 t}}\right]^{3/2} \exp\left[-\frac{m|\vec{u} - \vec{u}_0 e^{-\beta_0 t}|^2}{2kT\left(1 - e^{-2\beta_0 t}\right)}\right] \tag{1.83}$$

So at $t \to \infty$ a Maxwell distribution is obtained

$$W(\vec{u}, t; \vec{u}_0) \xrightarrow[t \to \infty]{} \left(\frac{m}{2\pi kT}\right)^{3/2} \exp\left[-m\frac{|\vec{u}|^2}{2kT}\right]$$

It appears that with the assumption introduced concerning $\omega[\vec{B}(\Delta t)]$, it is possible to obtain the distribution of a Brownian particle position $\vec{r}$ at time t if at $t = 0$ the particle was in point $\vec{r} = \vec{r}_0$ having velocity $\vec{u} = \vec{u}_0$. Indeed, since

$$\vec{r} - \vec{r}_0 = \int_0^t \vec{u}(t)\,dt$$

a number of transformations taking into account (1.79) brings to

$$\vec{R} = \vec{r} - \vec{r}_0 - \beta_0^{-1}\vec{u}_0\left(1 - e^{-\beta_0 t}\right) = \beta_0^{-1}\int_0^t \left[1 - e^{\beta_0(\xi - t)}\right]\vec{A}(\xi)\,d\xi \tag{1.84}$$

As for any value $\vec{R}$ determined by expression

$$\vec{R} = \int_0^t \psi(\xi)\vec{A}(\xi)\,d\xi \tag{1.85}$$

the probability distribution [19] has the form

$$W(\vec{R}) = \left[4\pi q \int_0^t \psi^2(\xi)\,d\xi\right]^{-3/2} \exp\left(-|\vec{R}|^2 / 4q \int_0^t \psi^2(\xi)\,d\xi\right) \tag{1.86}$$

we can find from (1.84) – (1.86)

$$\psi(\xi) = \beta_0^{-1}\left[1 - e^{\beta_0(\xi - t)}\right]$$

$$W(\vec{r}, t;\ \vec{r}_0, \vec{u}_0) = \left\{\frac{m\beta_0^2}{2\pi kT\left[2\beta_0 t - 3 + 4e^{-\beta_0 t} - e^{-2\beta_0 t}\right]}\right\}^{3/2} \times$$

$$\times \exp\left\{-\frac{m\beta_0^2\left[\vec{r} - \vec{r}_0 - \vec{u}_0\left(1 - e^{-\beta_0 t}\right)\beta_0^{-1}\right]^2}{2kT\left[2\beta_0 t - 3 + 4e^{-\beta_0 t} - e^{-2\beta_0 t}\right]}\right\} \tag{1.87}$$

For time periods larger than β_0^{-1} the stationary limit is then

$$W(\vec{r}, t;\ \vec{r}_0, \vec{u}_0) \simeq (4\pi D t)^{-3/2} \exp\left(-\frac{|\vec{r} - \vec{r}_0|^2}{4Dt}\right) \tag{1.88}$$

and for the introduced diffusion coefficient D

$$D = kT(m\beta_0)^{-1} = kT(6\pi a\eta)^{-1} \tag{1.89}$$

The formula for mean-square displacement resulting from (1.87) – (1.89) is

$$\left\langle |\vec{r} - \vec{r}_0|^2 \right\rangle = \frac{|\vec{u}_0|^2}{\beta_0^2}\left(1 - e^{-\beta_0 t}\right)^2 + $$

$$+ 3\frac{kT}{m\beta_0^2}\left(2\beta_0 t - 3 + 4e^{-\beta_0 t} - e^{-2\beta_0 t}\right) \tag{1.90}$$

Averaging (1.90) over all values $\vec{u}_0$ taking into account $\langle |\vec{u}_0|^2 \rangle = 3kT/m$ derives expression

$$\left\langle\left\langle |\vec{r} - \vec{r}_0|^2 \right\rangle\right\rangle = 6\frac{kT}{m\beta_0^2}\left(\beta_0 t - 1 + e^{-\beta_0 t}\right) \tag{1.91}$$

which at $t \to \infty$ leads to Einstein's result

$$\left\langle\left\langle |\vec{r} - \vec{r}_0|^2 \right\rangle\right\rangle = 6Dt = \frac{kT}{\pi a \eta}\, t \tag{1.92}$$

At the same time at $t \to 0$ it follows from (1.91) that

$$\left\langle\left\langle |\vec{r} - \vec{r}_0|^2 \right\rangle\right\rangle = 3\frac{kT}{m}t^2 = \left\langle |\vec{u}_0|^2 \right\rangle t^2 \tag{1.93}$$

Dependencies of types (1.92) and (1.93) will be discussed in Section 1.9.

Similarly it is possible to find the distributions defining probability $W(\vec{r},\vec{u},t;\ \vec{r}_0,\vec{u}_0)$ of simultaneously detecting position $\vec{r}$ and velocity $\vec{u}$ of a Brownian particle at time t, if at $t = 0$ the particle was located in the neighborhood of point $\vec{r} = \vec{r}_0$. Having rate $\vec{u} = \vec{u}_0$. Reference [19] gives the corresponding laws in the case where force field $\vec{F}(\vec{r},t)$ is present in the system.

The Langevin method briefly discussed here, which was first developed for solving problems of Brownian motion, can be successfully applied to describe a wide class of random processes (see Section 1.9) non-Markovian processes included.

1.8.2 The Kramers-Chandrasekhar Equation

Let us assume that the Brownian motion can be presented as a Markov process, i.e., the behavior of a Brownian particle at a given time moment is dependent only on the momentary values of the physical parameters and independent of the system history. In this sense the Markovian process can be shown as the "gradual development of transition probability" [19] and the probability distribution $W(\vec{r},\vec{u},t + \Delta t)$ in phase space for time $t + \Delta t$ can be deduced from distribution $W(\vec{r},\vec{u},t)$ for the previous time moment t using integral equation [19]

$$W(\vec{r},\vec{u},t + \Delta t) = \iint W(\vec{r} - \Delta\vec{r},\vec{u} - \Delta\vec{u},t)\ \times$$

$$\times\ \psi(\vec{r} - \Delta\vec{r},\vec{u} - \Delta\vec{u};\ \Delta\vec{r},\Delta\vec{u})\,d(\Delta\vec{r})\,d(\Delta\vec{u}) \tag{1.94}$$

As has already been mentioned, the expressions for increments $\Delta\vec{r}$ and $\Delta\vec{u}$ in time Δt are

$$\Delta\vec{r} = \vec{u}\Delta t, \qquad \Delta\vec{u} = -(\beta_0\vec{u} - \vec{F})\Delta t + \vec{B}(\Delta t) \tag{1.95}$$

where $\vec{r}$ and $\vec{u}$ are the radius-vectors of the coordinate and velocity of a Brownian particle respectively.

Let us now introduce such a function $\psi(\vec{r},\vec{u};\ \Delta\vec{u})$—the transition probability in velocity space that*

$$\Psi(\vec{r},\vec{u};\ \Delta\vec{r},\Delta\vec{u}) = \Psi(\vec{r},\vec{u};\ \Delta\vec{u})\,\delta(\Delta\vec{r} - \vec{u}\Delta t)$$

Then Equation (1.94) can be transformed into

$$W(\vec{r},\vec{u},t + \Delta t) = \int W(\vec{r} - \vec{u}\Delta t,\vec{u} - \Delta\vec{u},t) \times$$
$$\times\ \psi(\vec{r} - \vec{u}\Delta t,\vec{u} - \Delta\vec{u};\ \Delta\vec{u})d(\Delta\vec{u}) \tag{1.96}$$

Analogously it can be written

$$W(\vec{r} + \vec{u}\Delta t,\vec{u},t + \Delta t) = \int W(\vec{r},\vec{u} - \Delta\vec{u},\Delta t)\psi(\vec{r},\vec{u} - \Delta\vec{u};\ \Delta\vec{u})d(\Delta\vec{u}) \tag{1.97}$$

Expanding functions in Equations (1.96) and (1.97) using the Taylor series, one can obtain [19]

$$\left(\frac{\partial W}{\partial t} + \vec{u}\nabla_{\vec{r}}W\right)\Delta t + 0(\Delta t^2) =$$
$$= -\sum_i \frac{\partial}{\partial u_i}[W<\Delta u_i>] + \frac{1}{2}\sum_i \frac{\partial^2}{\partial u_i^2}[W<\Delta u_i^2>] +$$
$$+\sum_{i<j} \frac{\partial^2}{\partial u_i \partial u_j}[W<\Delta u_i \Delta u_j>] + 0[<\Delta u_i \Delta u_j \Delta u_k>] \tag{1.98}$$

where

$$<\Delta u_i^2> \equiv \int_{-\infty}^{+\infty} \Delta u_i \psi(\vec{r},\vec{u};\ \Delta\vec{u})d(\Delta\vec{u})$$

---

*We introduce notation $\delta(\vec{R}) = \delta(R_x)\delta(R_y)\delta(R_z)$

$$\langle \Delta u_i^2 \rangle \equiv \int_{-\infty}^{+\infty} \Delta u_i^2 \psi(\vec{r},\vec{u};\ \Delta\vec{u})\, d(\Delta\vec{u})$$

$$\langle \Delta u_i \Delta u_j \rangle \equiv \int_{-\infty}^{+\infty} \Delta u_i \Delta u_j \psi(\vec{r},\vec{u};\ \Delta\vec{u})\, d(\Delta\vec{u}) \tag{1.99}$$

Taking into account (1.82) and (1.95) the transition probability $\psi(\vec{r},\vec{u};\ \Delta u)$ can be presented in the form

$$\psi(\vec{r},\vec{u};\ \Delta\vec{u}) = (4\pi q \Delta t)^{-3/2} \exp\left[-\frac{|\Delta\vec{u} + (\beta_0\vec{u} - \vec{F})\Delta t|^2}{4q\Delta t}\right] \tag{1.100}$$

Obviously in this case

$$\langle \Delta u_i \rangle = -(\beta_0 u_i - F_i)\Delta t; \qquad \langle \Delta u_i^2 \rangle = 2q\Delta t + 0(\Delta t^2)$$

$$\langle \Delta u_i \Delta u_j \rangle = 0(\Delta t^2), \qquad q = \frac{\beta_0 kT}{m} = \beta_0^2 D \tag{1.101}$$

So Equation (1.98) is simplified and assumes the form (after proceeding to limit $\Delta t \to 0$)

$$\partial W/\partial t + \vec{u}\nabla_{\vec{r}} W + \vec{F}\,\nabla_{\vec{u}} W = \beta_0 \operatorname{div}_{\vec{u}}(W\vec{u}) + q\nabla_{\vec{u}}^2 W \tag{1.102}$$

This equation (sometimes called the Kramers-Chandrasekhar equation) is the most general form of the Fokker-Planck equation in phase space. At the same time this equation may be treated as a generalization of the Liouville theorem of classical dynamics for the case of Brownian motion. Indeed, the left part of (1.102) contains Stokes convective derivative D/Dt, acting upon W, whereas the components of the right part appeared due to consideration of the stochastic character of Brownian particle motion.

1.8.3 Telegrapher's Equation

Here we derived Equation (1.102) proceeding from the stochastic Langevin equation since his analysis provides an understanding of the approximations which are actually employed but not always stipulated in the consideration of diffusion kinetics problems or in the DCR theory. Let us rewrite Equation (1.102) for the one-dimensional case

$$\frac{\partial W}{\partial t} + u\frac{\partial W}{\partial x} = \frac{\partial}{\partial u}[(\beta_0 u - F(x))W] + \frac{kT\beta_0}{m}\frac{\partial^2 W}{\partial u^2} \tag{1.103}$$

and introduce the density probability function $\rho(x,t)$ for finding a Brownian particle in some point of space

$$\rho(x,t) = \int_{-\infty}^{+\infty} W(x,u,t)\,du \tag{1.104}$$

Function $\rho(x,t)$ in diffusion kinetics is described by the Smoluchowski Equation [66–69]

$$\frac{\partial \rho}{\partial t} + \frac{\partial}{\partial x}\left[\frac{F(x)}{\beta_0}\,\rho\right] = \frac{kT}{m\beta_0}\,\frac{\partial^2 \rho}{\partial x^2} \tag{1.105}$$

Now let us show that this equation can be derived from (1.103) after a series of simplifications [66]. Let us first note that integrating (1.103) over velocity space gives a continuity equation which expresses the balance on the density of a substance in the system

$$\frac{\partial \rho}{\partial t} + \frac{\partial}{\partial x}\,\rho\bar{u} = 0 \tag{1.106}$$

where

$$\bar{u} = \frac{1}{\rho}\int_{-\infty}^{+\infty} uW(x,u,t)\,du \tag{1.107}$$

After multiplying both parts of Equation (1.103) by u and integrating with respect to velocity space one can obtain

$$\frac{\partial}{\partial t}\,(\rho\bar{u}) + \frac{\partial}{\partial x}\,[\rho\langle u^2\rangle] + \beta_0\rho\bar{u} = F(x)\rho \tag{1.108}$$

where

$$\langle u^2\rangle \equiv \frac{1}{\rho}\int_{-\infty}^{+\infty} u^2 W(x,u,t)\,du \tag{1.109}$$

Equation (1.108) expresses the balance on the kinetic momentum in the system. After excluding $\rho\bar{u}$ from (1.106) and (1.108) an equation in configurational space is obtained

$$\frac{1}{\rho_0}\,\frac{\partial^2 \rho}{\partial t^2} + \frac{\partial \rho}{\partial t} + \frac{\partial}{\partial x}\left[\frac{F(x)\rho}{\beta_0}\right] = \frac{\partial^2}{\partial x^2}\left[\frac{\rho\langle u^2\rangle}{\beta_0}\right] \tag{1.110}$$

At long times ($t \gg \beta_0^{-1}$) or, which is the same, at $t > 0$, if $\beta_0 \to \infty$ one has

$$\rho\left[<u^2> - \frac{kT}{m}\right] \approx 0\left(e^{-2\beta_0 t}\right)$$

so that

$$\lim_{\beta_0 \to \infty} \frac{<u^2>}{\beta_0} = \frac{kT}{m\beta_0} = \frac{kT}{6\pi a\eta} = D \qquad (1.111)$$

and Equation (1.110) can be considered as a diffusion equation, but with an additional term $\beta_0^{-1}\partial^2\rho/\partial t^2$. It should also be required that

$$\lim_{\beta_0 \to \infty} [F(x)/\beta_0] = \nu(x) \qquad (1.112)$$

where $V(x)$ is the "drift" or migrational velocity of a Brownian particle.

If $|\partial^2\rho/\partial t^2| < \infty$, then at $\beta_0 \to \infty$ component $\beta_0^{-1}\partial^2\rho/\partial t^2$ can be ignored in (1.110) and thus Equation (1.105) is obtained.

Equation (1.110) is called the telegrapher's equation. Being a hyperbolic-type equation, it describes the situation more adequately at short times as opposed to the hyperbolic-type canonical diffusion equation. This more precise description is justified by Equation (1.110) by having the parameter of particle rate, whereas the usual diffusion equation lacks it.

It is obvious that each kinetics equation can be connected to a definite conservation law: any equation reflects the balance on some quantity which can be expressed through the moment of the distribution function. An ordinary diffusion equation implies only the conservation of the full number of particles, whereas Equation (1.110), as has already been mentioned, also implies that the conservation of their momentum is taken into account. A more detailed description of Brownian motion requires exceeding the limits of Equation (1.110) taking into account acceleration of the moving particles.

Now let us once more turn to a comparison of Equations (1.110) and (1.105). We shall consider only the simplest case: $F(x) = 0$, $\bar{u}(0) = 0$, $\rho(x,0) = \delta(x)$, $<u^2(0)> = kT/m$. In other words, we shall assume that the Brownian particles initially located at the origin of the coordinates are in equilibrium by kinetic energy with the environment, their initial velocities being characterized by symmetrical distribution and the particles not subject to external forces. In this case the solution of (1.105) at $\rho(x,0) = \delta(x)$ has the form

$$\rho_1(x,t) = \left(4\pi\frac{kT}{m\beta_0}t\right)^{-1/2} \exp\left[-x^2\Big/ 2\left(\frac{kT}{m\beta_0}t\right)\right] \qquad (1.113)$$

Let us also write the solution of (1.110) at $\rho(x,0) = \delta(x)$ and $\partial\rho/\partial t|_{t=0} = 0$:

$$\rho_2(x,t) = \begin{cases} \frac{1}{4}\exp\left(-\frac{1}{2}\rho_0 t\right)\beta_0\left[I_0(X) + \right. \\ \left. + \frac{\beta_0 t}{2} I_1(X)X^{-1}\right], & \text{if } |x| < (kT/m)^{1/2}t \\ 0, & \text{if } |x| > (kT/m)^{1/2}t \\ \frac{1}{2}\exp\left(-\frac{1}{2}\beta_0 t\right)\beta_0, & \text{if } x = \pm (kT/m)^{1/2}t \end{cases}$$

$$X = \frac{1}{2}\beta_0(t^2 - mx^2/kT)^{1/2} \tag{1.114}$$

where $I_\nu(z)$ is a Bessel function of an imaginary argument and of order ν.

Evidently at $t \gg \beta_0^{-1}$ function $\rho_2(x,t)$ tends asymptotically to $\rho_1(x,t)$ as can be expected since Equations (1.105) and (1.110) have the same stationary solution.

A comparison of (1.105) and (1.110) shows why a diffusion equation is inadequate at short times $t \lesssim \beta_0^{-1}$. According to (1.105) the diffusing particles located at the origin of the coordinates at $t = 0$ at $t > 0$ have a finite (though small) probability of being detected at any point no matter how far from the initial point. However, it is evident that within minimally short time t they cannot move to a distance larger than value $|x| = (kT/m)^{1/2}t$ (determined by the rate at the initial time moment). This is adequately shown in the solution of (1.110). It should be noted that taking into account these "delay effects" is, as a rule, unnecessary: particles formally moving at an "infinite rate" usually make too small a relative contribution. An example of an exception is the analysis of fast spreading processes of the chemical transformation front in condensed phases (in particular, new phenomenon investigated in [70,71] in the low-temperature chemistry of solids—the appearance of a common wave for both the destruction and the chemical transformation which propagates along the frozen reagent mixture).

It has already been mentioned that Equations (1.105) and (1.110) do not exhaust all the possible relations ("conservation laws") between moments of the phase space distribution function of a system. It is well known that in the most general form kinetics equations are integral-differential, i.e., contain an infinite number of functional relations [72]. However, in certain cases it is possible to give a substantiated expansion of the integral operators of such general kinetics "master equations" [73] into a converging series in partial coordinate derivatives.

1.8.4 Expansion of an Integral Operator of a Master Kinetic Equation into a Series in Differential Operators [73]

Let 'a' be the fluctuating quantity, $P(a, t)$ the function of probability distribution at time t. For the Markovian process, $P(a,t)$ satisfies the equation

$$\partial P(a,t)/\partial t = \int [W(a|a')P(a',t) - W(a'|a)P(a,t)]da' \tag{1.115}$$

where $W(a|a')$ is the probability of a transition from state a' into state 'a' per time unit. Equation (1.115) is equivalent to the relation

$$\partial P(a,t)/\partial t = \sum_{n=1}^{\infty} \frac{1}{n!} \left(-\frac{\partial}{\partial a}\right) \alpha_n(a)P(a,t) \tag{1.116}$$

where

$$\alpha_n(a) = \int (a' - a)^n W(a'|a)da'$$

Equation (1.116) at $\alpha_k = 0$ $(k > 2)$ is a Fokker-Planck equation. However, the procedure of formal discarding of members containing derivatives of higher than second order (1.116) is not always correct. Nevertheless, it is possible to formulate a more or less common method which provides for the derivation of asymptotic solutions (1.116) for a wide class of functions. In some respects the essence of the method is analogous to known asymptotic quasiclassical methods (obtaining the Hamilton-Jacobi equation from the Schrödinger equation; the eikonal equation in geometric optics from the Helmholtz wave equation, etc.).

Let function $W(a|a')$ be such that it can be presented in the form

$$W(a|a') = \phi(a'/\Omega;\ \Delta a) \tag{1.117}$$

where $\Delta a = a - a'$, Ω is the characteristic "volume" for quantity a (henceforth $\Omega \to \infty$).

For example if a is the number of particles, Ω is the volume occupied by these particles. Then $\alpha_n = \alpha_n(a/\Omega)$.

Let us replace variables in Equation (1.116). For this purpose, we first assume that

$$a = \Omega\varphi(t) + \Omega^{1/2}x \tag{1.118}$$

where x is a new variable; $\varphi(t)$ is an arbitrary time function. Substituting (1.118) into (1.116) we obtain

$$\frac{\partial P(x,t)}{\partial t} - \Omega^{1/2}\varphi'(t)\frac{\partial P(x,t)}{\partial x} =$$

$$= \sum_{n=1}^{\infty} \frac{\Omega^{-n/2}}{n!} \left(-\frac{\partial}{\partial x}\right)^n \alpha_n[\varphi(t) + \Omega^{-1/2}x]P(x,t) \tag{1.119}$$

The second term of the left part in (1.119) disappears if function $\varphi(t)$ obeys the calibrating relationship $\varphi'(t) = \Omega^{-1}\alpha_1[\varphi(t)]$. Let us also introduce a new time scale $\tau = t\Omega^{-1}$ corresponding to the increase of relaxation time of quantity 'a' with the increase of its "volume". Then if we expand the right part of (1.119) as a series in power of $\Omega^{-1/2}$ and regroup the terms of the series we obtain

$$\frac{\partial P(x,\tau)}{\partial \tau} = \sum_{m=2}^{\infty}\left[\Omega^{(m-2)/2}\, m!\right]^{-1} \sum_{n=1}^{m} \binom{m}{n} \alpha_n^{(m-n)} \times$$

$$\times\ [\varphi(\tau)] \left(-\frac{\partial}{\partial x}\right)^n x^{m-n}\, P \tag{1.120}$$

An equation for moments $\langle x^k \rangle$ of value x of order k following from (1.120) is

$$\frac{d}{d\tau}\langle x^k\rangle = \sum_{m=2}^{\infty}\left[\Omega^{(m-2)/2}\, m!\right]^{-1} \sum_{n=1}^{m+k} \binom{m}{n} \alpha_n^{(m-n)} \times$$

$$\times\ [\varphi(\tau)]\ \frac{k!}{(k-n)!} \left\langle x^{m+k-2n} \right\rangle$$

Relationship (1.120) in zero order in $\Omega^{-1/2}$ gives

$$\frac{\partial P}{\partial \tau} = -\alpha_1[\varphi(\tau)]\, \frac{\partial}{\partial x}\, xP + \frac{1}{2}\, \alpha_2[\varphi(\tau)]\, \frac{\partial^2 P}{\partial x^2} \tag{1.121}$$

In describing fluctuations in the equilibrium state, value $\varphi(\tau)$ can be replaced by its equilibrium value $\varphi(\infty)$. In this case, the solution of Cauchy problem for Equation (1.121) at $P(x,0) = \delta(x - x_0)$ assumes the form

$$P(x,\tau) = \left[\frac{\pi\alpha_2}{(-\alpha_1')}\left(1 - e^{\alpha_1'\tau}\right)\right]^{-1/2} \exp\left\{-\left(x - x_0 e^{\alpha_1'\tau}\right)^2 \Big/ \left[\left(-\frac{\alpha_2}{a_1'}\right)\left(1 - e^{2\alpha_1'\tau}\right)\right]\right\}$$

If the component containing $\Omega^{-1/2}$ is taken into account in the expansion of (1.120), the following additional term appears in the right part of Equation (1.121)

$$\frac{1}{3!}\ \Omega^{-1/2}\left[-3\alpha_1''\, \frac{\partial}{\partial x}\, x^2 P + 3\alpha_2'\, \frac{\partial^2}{\partial x^2}\, xP - \alpha_3\, \frac{\partial^3 P}{\alpha x^3}\right]$$

Let us consider the one-dimensional motion of a heavy particle of mass M in the gas of light particles of mass m, $m \ll M$ as an example illustrating an application of the method developed. In this case probability

$W(v'|v)$ for the velocity of the heavy particle to change is given by expression

$$W(v'|v) = \nu\left(\frac{M+m}{2m}\right)^2 |v' - v| f\left(\frac{M+m}{2m}v' - \frac{M-m}{2m}v\right)$$

where ν is the number of light particles per length unit, $f(u)$ is the Maxwell distribution of the velocity of these particles.

Assuming that

$$a = \frac{M+m}{m}v \quad \text{and} \quad \Omega = \frac{M+m}{m}$$

we obtain

$$W\left(\frac{a'}{\Omega} \,\middle|\, \frac{a}{\Omega}\right) = \frac{\nu}{4}\,\Omega|a' - a|\, f\left(\frac{a}{\Omega} + \frac{a' - a}{2}\right)$$

In the zero order approximation in Ω we have a Fokker-Planck Equation (1.121) where

$$\alpha_1{}' = -\frac{m}{2kT}, \qquad \alpha_2 = 4\nu\left(\frac{2kT}{m}\right)^{1/2}$$

Replacing variables $v = x\Omega^{-1/2}$, $t = \Omega\tau$ we obtain the well-known diffusion equation in velocity space

$$\frac{\partial P(v,t)}{\partial t} = \frac{4\nu m}{M}\left(\frac{2kT}{\pi m}\right)^{1/2} \frac{\partial}{\partial v}\left(vP + \frac{kT}{M}\frac{\partial P}{\partial v}\right)$$

which is used for investigating relaxation kinetics in gases [74].

Unfortunately the above-mentioned expansions of integral operators cannot always be substantiated, so that in analyzing DCR kinetics one has to consider only equations in partial derivatives of second order without sufficient evaluation of the discarded components. Obviously the limitation by partial derivatives to second order (diffusion equation) permits describing only slow enough processes of "concentrational relaxation". The genesis of such a slow (diffusional) relaxation becomes more understandable if one proceeds to discrete models of random walks. Considering stochastic walks on the energy levels of the quantum oscillator placed in a thermostat, it can be shown (see for example [5]) that taking into account transitions between adjacent energy levels corresponds to the diffusion limit. Introducing into consideration transitions over two energy levels in the continuous limit leads to an equation with partial derivatives of the III order. Taking into account energy "jumps" to higher or lower (by three and more) energy levels determines the transition to equations in partial derivatives of higher orders. Evidently the consideration of random walks in configurational space will lead to the same result: allowing for jumps onto the second and farther coordination spheres in the model of random walks gives rise to the appearance of partial derivatives of third and higher coordinate orders in the equation.

Thus it becomes clear that in order to describe the concentrational relaxation kinetics of a strongly non-equilibrium system, it is insufficient to use a diffusion equation. In particular, this concerns systems with large initial density fluctuations, when "the probabilities of discrete changes in the system state" in phase space differ from the corresponding probabilities in the case of small gradient relaxation. Exceeding the limits of diffusion approximations could consist of an analysis of equations with partial derivatives of the third order and mixed ones (coordinate and time derivatives). Obviously to adequately represent the master kinetic equation with partial derivatives of higher orders it is necessary to devise universal techniques of expanding the integral operator into series in differential ones.

Proceeding to higher order equations makes their solution considerably more complicated. This is why quite often instead of increasing the order of an equation, one prefers a complication of the diffusion equation, for example, by introducing phenomenological dependences of diffusion coefficients D on coordinates or concentrations c. The necessity of introducing dependences $D(c)$ at large diffusing particle concentrations is formally substantiated by the particle effect on the matrix in which they are dissolved. In many cases these assumptions are completely justified, for example in considering diffusion transfer of a number of gas molecules (ammonia, carbon dioxide) in some polymer materials when the effects of matrix plasticization are possible. However, in not all introduction cases is the influence of diffusing particles upon the system evident, and thus the question of the inadequacy of the true diffusion approximation with respect to the problem considered may arise. This is why clarification of the genesis of dependence $D(c)$ is of particular interest in cases when this dependence should necessarily be introduced for a formal description of the observed regularities.

And finally, as has been stated above, it has proven insufficient to use both diffusion approximation as well as approximations of a higher order if the fluctuation values are large to describe chemical transformation kinetics taking into account initial fluctuations in reagent distribution. Nevertheless, recently a quite effective approximation method was worked out. It will be discussed in Section 3.7.

1.9 DIFFUSION MOTION OF QUANTUM PARTICLES IN A MEDIUM

"Classical" DCR theory is without doubt useful in the systematization of experimental data, but modelling dynamics via Brownian motion calls forth serious objections [75] which boil down to this.

1. The nature of the migration (diffusion mobility) remains unclear. Various interparticle (or other) interactions and the energy exchange of migrating particles with a medium may dramatically influence the character of particle migration (in particular, simplify it considerably); that is why

it is impossible to reproduce various dynamic aspects of the problem with the aid of only one parameter—diffusion activation energy.

2. The assumption about the diffusional character of the migration is equivalent to the statement that mean-square shift $\langle R^2\rangle_t$ is proportional to time t within long times and the equation

$$D = \lim_{t\to\infty} \langle R^2\rangle_t/2dt \tag{1.122}$$

(d—dimensionality of the problem), determining diffusion coefficient holds. However, there are several theoretical arguments and experimental data that contradict this statement. In particular, it does not hold for the migration of light atoms sorpted on a crystal surface, for example atoms of ^3He and ^4He on a graphite substrate [76]. In this case, coherent motion can make an appreciable contribution to surface migration exceeding those of a diffusional character.

An important aspect of the problem is elucidating the influence of the degree of order of the medium itself upon the character of the stochastic particle motions.

Authors of paper [77], while discussing this problem in connection with experiments on fluorescence quenching in a diluted solution of chlorophyll, discovered that for solutions (which are not periodic structures) the dependence $\langle R^2\rangle_t \sim t^{5/6}$ should be realized instead of (1.122). Thus, in the general case, particle migration can be expected to carry both a diffusional and non-diffusional character. This necessitates more general approaches to analyzing particle dynamics and in particular to seeking the dependence of the mean-square shift on time.

Specifying the transfer mechanisms (diffusion mobility) requires introducing into explicit consideration the interaction Hamiltonian $\mathcal{H}$. In describing the dynamics of the system it is convenient to use equations of motion, written with the help of density matrix ρ, since it provides for a simultaneous calculation of both the averaged over statistic ensemble and the quantum mechanical matrix elements. It is convenient to consider the multiparticle problem employing the secondary quantization method using representation of occupation numbers [78] for operators and wave functions.

1.9.1 Model Hamiltonians (Representation of Occupation Numbers)

The Hamiltonian of the investigated system in the representation of secondary quantization can be obtained using the correspondence principle for describing quantum operators instead of classical quantities [79].

Let us see how this is done. Let us assume that we know the solution of a one-particle wave equation

$$\hat{\mathcal{H}}\varphi_j(x) = \varepsilon_j\varphi_j(x) \tag{1.123}$$

Funtions $\varphi_j(x)$ form a complete orthonormalized basis set such that the vector of an arbitrary state $\psi(x)$ of the system can be expanded in these functions.

On the basis of (1.123) we shall introduce field operators

$$\psi(x) = \sum_j c_j \varphi_j(x), \qquad \text{and} \qquad \psi^+(x) = \sum_j c_j^+ \varphi_j{}^*(x) \tag{1.124}$$

where c_j, c_j^+ are operators of particle annihilation and creation in state j.

With the help of the introduced field operators, one can write the Hamiltonian of the system as the secondary quantization representation. According to general procedure [79], this requires writing an expression for the mean value of the operator through separate particle wave functions and then replacing the wave functions by a field operator. For the one-particle operator of kinetic energy $\hat{T}$, for example, the mean value when using functions (1.123) has the form

$$\langle\hat{T}\rangle = \int \psi^*(x)\hat{T}\,\psi(x)dx$$

Then in the secondary quantization representation

$$\langle\hat{T}\rangle = \int \psi^*(x)\hat{T}\,\psi(x)\;dx = \sum_{j,e} c_j c_l{}^+ T_j{}^l$$

where $T_j{}^l = \int\varphi^*\hat{T}\varphi_j dx$ is the matrix element of operator $\hat{T}$.

To define the properties of operators c_j, c_j^+ it is convenient to proceed to the bracket notations of Dirac [78], assuming $\varphi_j(x) = \langle x|j\rangle$, i.e., introducing state vectors "ket" $|j\rangle$ and "bra" $\langle x|$. Field operators and operators c_j act upon state vector Φ of the multiparticle system which can be written in the form

$$\Phi = |n_1, n_2, \ldots, n_k, \ldots, n_N\rangle$$

where n_k is the number of particles on the level referring to wave function φk.

Acting upon the state vector Φ, operators c_j and c_j^+ decrease or increase the number of particles correspondingly on level j. For bosons, for example

$$c_j|\ldots n_j \ldots\rangle = \sqrt{n_j}\,|\ldots n_{j-1}\ldots\rangle$$

$$c_j^+|\ldots n_j \ldots\rangle = \sqrt{n_{j+1}}|\ldots n_{j+1}\ldots\rangle$$

This is the reason operators c_j^+ and c_j are referred to as operators of creation or annihilation. Their properties are assigned by commutation relation

$$c_j c_k{}^+ \mp c_k{}^+ c_j = \delta_{jk}$$

Here the sign "−" refers to boson operators, the sign "+" to fermion ones. For both types of statistics, the operator of the particles number has the same form $n_j = c_j^+ c_j$.

Let us present several types of Hamiltonians in the secondary quantization representation which will be required later.

A Hamiltonian of the following type is used for the simplest two-level problem in [80].

$$\hat{\mathscr{H}} = \gamma\{|1><2| + |2><1|\} = \gamma\{c_1^+ c_2 + c_2^+ c_1\} \tag{1.125}$$

It describes only transitions between levels, parameter γ characterizing the corresponding transition matrix element. Hamiltonian (1.125) can be easily generalized for describing transitions in a one-dimensional chain

$$\hat{\mathscr{H}} = \frac{1}{2} \sum_{m,n}^{\infty} \tilde{\beta}_{mn}^{(0)} \{|m><n| + |n><m|\} \tag{1.126}$$

Here parameters $\tilde{\beta}_{mn}^{(0)}$ depend only on difference $(m - n)$ and describe the free motion of a particle conserving a quasi-momentum. In the case of a uniform chain $\tilde{\beta}_{mn}^{(0)} = \gamma$.

In investigating the nature of diffusion mobility (migration) interactions can be of various types. A significant role is played by model Hamiltonians, which include important features of the problem, and permit its exact solution. Let us consider one of these model Hamiltonians used in [81] for the analysis of atom migration on crystal surfaces. It has the form (assuming $h/2\pi = 1$)

$$\hat{\mathscr{H}} = E_0 \sum_n a_n^+ a_n + \Omega c^+ c + \sum_\lambda \omega_\lambda b_\lambda^+ b_\lambda + \sum_{m \neq n} J_{nm} a_n^+ a_m +$$

$$+ \sum_{n \neq m} \varphi_{nm} a_n^+ a_m (c + c^+) + \sum_{n,\lambda} \omega_\lambda X_\lambda^n a_n^+ a_n (b_\lambda + b_{-\lambda}^+) +$$

$$+ \sum_{n,\lambda} Y_\lambda^n a_n^+ a_n (c + c^+)(b_\lambda + b_{-\lambda}^+)$$

where a_n^+ and a_n—operators of creation and annihilation of an adsorbed atom in an n-th site of the lattice; terms $E_0 \Sigma_n a_n^+ a_n$ and $\Sigma_{n \neq m} J_{mn} a_n^+ a_m$ describe the motion parallel to the plane of the surface, E_0 being the energy of the adsorbed atom in place n in its ground vibrational state; $|J_{mn}|^2$ refers to the probability of an atom transition from site n to site m due to coherent motion; c^+ and c—operators of creation and annihilation of adsorbed atom vibration in a direction perpendicular to the surface (term $\Omega c^+ c$ presents the vibration energy); b_λ^+ and b_λ—creation and annihilation operators of phonons with wave length λ and frequency ω_λ (the zone index is omitted for simplicity); term $\Sigma_\lambda \omega_\lambda b_\lambda^+ b_\lambda$ presents the lattice phonon energy; $\Sigma_{n \neq m} \varphi_{nm} a_n^+ a_m (c + c^+)$ corresponds to the vibrationally induced intersite transitions; terms $\Sigma_{n,\lambda} \omega_\lambda X_\lambda^n a_n^+ a_n (b_\lambda + b_{-\lambda}^+)$ and $\Sigma_{n,\lambda} Y_\lambda^n a_n^+ a_n (b_\lambda + b_{-\lambda}^+)(c + c^+)$ take into account change of energy E_0 resulting from the interaction of vibrations of the adsorbed atom and the lattice.

Coupling constants J_{nm}, φ_{nm}, X_λ^n, Y_λ^n can be calculated if the potential of the interaction between the adsorbed atom and lattice atoms is known near an equilibrium distance of the adsorbed atom from the surface at fixed equilibrium positions of lattice atoms.

Thus the postulated model Hamiltonian takes into account the motion of the adsorbed atom both parallel to the surface (in the periodic lattice potential) and vibrations perpendicular to the surface. These types of motion are connected since the distance from the surface changes with vibration excitation. The periodic potential determining the parallel motion also changes. Moreover, if the adsorbed atom strongly interacts with the surface it distorts the lattice around it, which in its turn affects the interaction potential of the atom with the lattice.

The Hamiltonian formally takes into account enough various interactions of the migrating atom. However, there are many parameters to be found. Its determination is quite difficult and problematic considering the complexity of the experimental investigations. Therefore in most cases the problem should be simplified by specifying the possible interactions. The introduction of concepts about the stochastic character of motion in a condensed phase can be most fruitful in this respect. This approach was suggested in [80] and is discussed below.

1.9.2 The Two-Level Problem

Let us use Hamiltonians built on the basis of (1.125) for a two-level system

$$\hat{\mathcal{H}} = \{|1\rangle\langle 2| + |2\rangle\langle 1|\}[\gamma + \tilde{\beta}(t)] + \alpha(t)|2\rangle\langle 2| \tag{1.127}$$

and for a one-dimensional chain

$$\hat{\mathcal{H}} = \frac{1}{2} \sum_{m,n=-\infty}^{\infty} [\tilde{\beta}_{mn}^{(0)} + \tilde{\beta}_{mn}(t)]\{|m\rangle\langle n| + |n\rangle\langle m|\} \tag{1.128}$$

Random values $\alpha(t)$, $\tilde{\beta}(t)$ and $\tilde{\beta}_{mn}(t)$ define the effect of a stochastic medium on the migrating particle. The statistical characteristics of the random process [$\tilde{\beta}(t)$ for example] are described by assigning its correlator, the form of which will be discussed below.

The motion of a particle (a quantum one in the general case) is described using density matrix $\hat{\rho}$. The operator of particle density (or density matrix) is built with the aid of field operators ψ (1.124):

$$\hat{\rho}(x) = \quad dx\psi^+(x')\delta(x - x')\psi(x) = \sum_{ij} c_i^+ c_j \varphi_i^*(x)\varphi_j(x) \tag{1.129}$$

The dependence of density matrix on time is defined by motion equations

$$i\frac{\partial\hat{\rho}}{\partial t} = -[\hat{\rho}, \hat{\mathscr{H}}] = -(\hat{\rho}\,\hat{\mathscr{H}} - \hat{\mathscr{H}}\,\hat{\rho}) \tag{1.130}$$

To illustrate the applied mathematical methods we shall consider a simple example of a two-level problem with a Hamiltonian (1.127).

Using Equation (1.130) we shall write the explicit form of an equation for matrix elements of density matrix

$$i\frac{d\rho_{11}}{dt} = (\gamma + \tilde{\beta})(\rho_{21} - \rho_{12}), \quad i\frac{d\rho_{22}}{dt} = (\gamma + \tilde{\beta})(\rho_{12} - \rho_{21})$$

$$i\frac{d\rho_{12}}{dt} = -\alpha\rho_{12} + (\gamma + \tilde{\beta})(\rho_{22} - \rho_{11}) \tag{1.131}$$

Density matrix ρ (or ρ_{ij}) of (1.131) depends on random quantities $\alpha(t)$ and $\beta(t)$. The derived equations should be averaged over the distribution of these quantities.

The statistical characteristics of value z are entirely described by function $p(z)$ called probability density (or distribution function)

$$p(z) > 0, \quad \int_{-\infty}^{+\infty} p(z)dz = 1$$

In particular the Gaussian random quantity is determined by probability density of the type

$$p(z) = \frac{1}{\sqrt{2\pi}\,\sigma}\exp(-z^2/2\sigma^2)$$

where σ is the distribution parameter.

Probability density function can be characterized by moments

$$M_n = \langle z^n \rangle = \int_{-\infty}^{+\infty} z^n p(z)dz$$

For Gaussian random quantity

$$M_1 = \langle z \rangle = 0, \quad M_2 = \langle z^2 \rangle = \sigma^2$$

and it can be easily shown that

$$M_{2n+1} = 0, \quad M_{2n} = (2n - 1)!!\,\sigma^{2n}$$

Quite important are mean values $\langle zf(z)\rangle$, where $f(z)$ is an arbitrary function. The mean value of function $f(z)$ itself is determined by the expression

$$\langle f(z)\rangle = \int_{-\infty}^{+\infty} f(z)p(z)dz$$

use of which makes it simple to obtain the relation for Gaussian quantity

$$\langle zf(z)\rangle = \frac{1}{\sqrt{2\pi}\,\sigma} \int_{-\infty}^{+\infty} zf(z)\exp(-z^2/2\sigma^2)dz \equiv \sigma^2 \left\langle \frac{df(z)}{dz} \right\rangle \tag{1.132}$$

This formula plays a significant role in the discussions following. It is easily generalized for multi-dimensional random vector quantity $\vec{z}$ with components z_i

$$\langle z_i f(\vec{z})\rangle = B_{ij} \left\langle \frac{\partial f(\vec{z})}{\partial z_j} \right\rangle \tag{1.133}$$

where

$$B_{ij} = \langle z_i z_j\rangle$$

Determining correlator B_{ij} presupposes the application of two-dimensional distribution function $p(z_1, z_2)$, so that

$$B_{ij} = \int z_1 z_2 p(z_1, z_2)dz_1 dz_2$$

The given definitions are easily generalized for random processes $z(t)$ and fields $z(\vec{x}, t)$. However when proceeding to random processes it is necessary to specify operations analogous to differentiation in (1.132) and (1.133). Since quantities $z(t)$ are functions themselves, consideration of random processes and fields leads to the operation of functional differentiation, the detailed account of which is given in monograph [82]. Let us write out the generalized relation (1.132) for random processes of the Gaussian type usually called the Furutzu-Novikov formula

$$\langle z(t)f[z(\tau)]\rangle = \int d\tau_1 B(t, \tau_1) \left\langle \frac{\delta f[z(\tau)]}{\delta z(\tau_1)} \right\rangle$$

Employing this relation requires specification of the type of correlator $B(t, t')$ defining the type of random processes. In solving physical problems, a special role is played by delta correlated processes, since such an approximation of parameter fluctuations has a clear physical nature and a closed equation [82] can be derived for the probability density of the corresponding dynamic systems.

In the general case, the correlation function of δ-correlated Gaussian process has the form

$$\langle z(t)z(t')\rangle = B(t)\,\delta(t - t') \qquad (\langle z\rangle = 0)$$

Let us consider a particular case, assuming B = const. for correlators of $\alpha(t)$ and $\tilde{\beta}(t)$ processes

$$\langle\tilde{\beta}(t)\tilde{\beta}(\tau)\rangle = 2\mu\,\delta(t - \tau), \qquad \langle\alpha(t)\alpha(\tau)\rangle = 2\delta\,\delta(t - \tau)$$

$$\langle\alpha(t)\rangle = \langle\tilde{\beta}(t)\rangle = \langle\alpha(t)\tilde{\beta}(\tau)\rangle = 0 \tag{1.134}$$

Now let us employ the given relationships for the analysis of Equations (1.131). Averaging the right and left parts of Equations (1.131) (over the distribution of $\alpha(t)$ and $\tilde{\beta}(t)$ quantities) we obtain the following system of equations for the density matrix:

$$\frac{d\langle\rho_{11}\rangle}{dt} = i\gamma(\langle\rho_{12}\rangle - \langle\rho_{21}\rangle) + i(\langle\tilde{\beta}\rho_{12}\rangle - \langle\tilde{\beta}\rho_{21}\rangle)$$

$$\frac{d\langle\rho_{22}\rangle}{dt} = i\gamma(\langle\rho_{21}\rangle - \langle\rho_{12}\rangle) + i(\langle\tilde{\beta}\rho_{21}\rangle - \langle\tilde{\beta}\rho_{12}\rangle)$$

$$\frac{d\langle\rho_{12}\rangle}{dt} = i\langle\alpha\rho_{12}\rangle + i\gamma(\langle\rho_{11}\rangle - \langle\rho_{22}\rangle) + i(\langle\tilde{\beta}\rho_{11}\rangle - \langle\tilde{\beta}\rho_{22}\rangle) \tag{1.135}$$

To obtain closed equations for the averaged density matrix $\langle\rho\rangle$, the mean values of type $\langle\tilde{\beta}\rho_{ik}\rangle$ and $\langle\alpha\rho_{ik}\rangle$ should be decoupled. This can be done using the Furutzu-Novikov formula for a multi-dimensional random process which can be presented in the form

$$\langle\tilde{\beta}\rho_{ik}\rangle = \sum_{\lambda=\alpha}^{\tilde{\beta}} \int_{-\infty}^{+\infty} d\tau\langle\tilde{\beta}(t)\lambda(\tau)\rangle \left\langle\frac{\partial\rho_{ik}(t)}{\partial\lambda(\tau)}\right\rangle \tag{1.136}$$

Since correlators $\langle\tilde{\beta}(t)\lambda(\tau)\rangle$ are proportional to $\delta(t - \tau)$ it is sufficient to know functional derivative $\delta\rho_{ik}(t)/\delta\lambda(\tau)$ at $t = \tau$. Under this condition functional derivatives can be expressed through the density matrix at the same time moment [82]. Following the method of calculation given in [82] one can obtain the expressions

$$\frac{\delta\rho_{11}(t)}{\delta\tilde{\beta}(t)} = i[\rho_{12}(t) - \rho_{21}(t)], \qquad \frac{\delta\rho_{22}(t)}{\delta\tilde{\beta}(t)} = i[\rho_{21}(t) - \rho_{12}(t)]$$

$$\frac{\delta\rho_{11}(t)}{\delta\alpha(t)} = \frac{\delta\rho_{22}(t)}{\delta\alpha(t)} = 0, \qquad \frac{\delta\rho_{12}(t)}{\delta\alpha(t)} = i\rho_{12}(t)$$

$$\frac{\delta\rho_{21}(t)}{\partial\alpha(t)} = -i\rho_{21}(t)$$

$$\frac{\delta\rho_{12}(t)}{\delta\tilde{\beta}(t)} = i[\rho_{11}(t) - \rho_{22}(t)], \qquad \frac{\delta\rho_{21}(t)}{\delta\tilde{\beta}(t)} = i[\rho_{22}(t) - \rho_{11}(t)] \tag{1.137}$$

Substituting (1.137) first into (1.136) then into (1.135) and taking into account (1.134) for averaged density matrix $<\rho_{ik}>$, we obtain a closed system of equations of the following type

$$\frac{d<\rho_{11}>}{dt} = -2\mu(<\rho_{11}> - <\rho_{21}>) + i\gamma(<\rho_{12}> - <\rho_{21}>)$$

$$\frac{d<\rho_{22}>}{dt} = -2\mu(<\rho_{22}> - <\rho_{11}>) + i\gamma(<\rho_{21}> - <\rho_{12}>)$$

$$\frac{d<\rho_{12}>}{dt} = -\delta<\rho_{12}> - 2\mu(<\rho_{12}> - <\rho_{21}>) + i\gamma(<\rho_{11}> + <\rho_{22}>) \tag{1.138}$$

For solving System (1.138) it is convenient to employ new variables

$$<\rho_{11}> = \frac{1}{2} + \rho, \qquad <\rho_{22}> = \frac{1}{2} - \rho,$$

$$<\rho_{12}> = R + iJ, \qquad <\rho_{21}> = R - iJ \tag{1.139}$$

transforming (1.138) into

$$d\rho/dt = -4\pi\rho - 2\gamma J, \qquad dJ/dt = -\delta J - 4\mu J + 2\gamma\rho$$

$$dR/dt = -\delta R \tag{1.140}$$

System (1.140) is easily solvable. Substituting its solution into (1.139) one finds averaged density matrix $<\rho_{ik}>$. As an example let us write out the expression for diagonal matrix element $<\rho_{11}>$:

$$<\rho_{11}> = \frac{1}{2} + \frac{1}{2}\exp\left(-4\mu t - \frac{\delta}{2}t\right)\left\{A\,\mathrm{Ch}\left[\left(\frac{\delta^2}{4} - 4\gamma^2\right)^{1/2}t\right] + B\,\mathrm{Sh}\left[\left(\frac{\delta^2}{4} - 4\gamma^2\right)^{1/2}t\right]\right\} \tag{1.141}$$

where A and B are constants, coefficient β being real at $\delta^2/4 > 4\gamma^2$.

It follows from Formula (1.141) that value $<\rho_{11}>$ tends to 1/2 at $t \to \infty$, the approaching to the limit is monotone if $\delta^2/4 > 4\gamma^2$, and of an oscillating character at $\delta^2/4 < 4\gamma^2$. The corresponding expressions for $<\rho_{22}>$ and $<\rho_{12}>$ show that $<\rho_{22}> \to 1/2$ and $<\rho_{12}> \to 0$ at $t \to \infty$; the time dependence having a monotone or oscillating character depending on the relation between parameters $\delta^2/4$ and $4\gamma^2$. Thus if there is a stochastic effect then with time the system reaches equilibrium distribution described by unit density matrix $\hat{\rho} = 1/2\ \hat{I}$.

1.9.3 Non-Stationary Problem of Wave Packet Spreading

Now let us consider a one-dimensional chain described by Hamiltonian (1.128) for analyzing the nonstationary wave packet spreading. Random functions $\tilde{\beta}_{mn}(t)$ shall be assigned by correlators of the type

$$<\tilde{\beta}_{mn}(t)> = 0$$

$$<\tilde{\beta}_{mn}(t)\tilde{\beta}_{m'n'}(t')> = 2G(m-n)\,\delta(t-t')(\delta_{mm'}\delta_{nn'} + \delta_{mn'}\delta_{m'n} - \delta_{mn}\delta_{m'n'}\delta_{nn'}) \qquad (1.142)$$

where $G(m-n)$ depends only on temperature and parameters of the crystal lattice. (Derivation of the expression for these quantities is given in [80]).

The choice of correlation functions in Formula (1.142) is determined by the fact that this choice provides for obtaining closed and quite simple equations for the density matrix averaged over all realizations of random process $\tilde{\beta}_{mn}(t)$.

The Shrődinger equation in the impulse representation for a system with Hamiltonian (1.128) assumes the form

$$i\frac{\partial\psi(K,t)}{\partial t} = \varepsilon_K\psi(K,t) + \int_{-\pi}^{\pi} R_t(K,q)\,\psi(q,t)\,dq \qquad (1.143)$$

where

$$\varepsilon_K = \sum_{m=-\infty}^{\infty} \tilde{\beta}^{(0)}_{n,m+n} e^{-ikm}$$

$$R_t(K,q) = \frac{1}{2\pi}\sum_{m,n=-\infty}^{\infty} e^{-ikn+iqm}\tilde{\beta}_{mn}(t); \qquad R_t(K,q) = R_t^{+}(q,K)$$

Using (1.143) one can obtain the following equation for the density matrix in momentum representation $\rho(K, K')$:

$$i \frac{\partial \rho(K,K')}{\partial t} = \int_{-\pi}^{\pi} dq \int_{-\pi}^{\pi} dq' B(K,K'|q,q')\rho(q,q') +$$

$$+ (\varepsilon_K - \varepsilon_{K'})\rho(K,K') \tag{1.144}$$

$$B(K,K'|q,q') = R_t(K,q)\delta(K' - q') - R_t(q',K')\delta(q - K)$$

Averaging the left and right parts of relation (1.144) we obtain an equation for density matrix $<\rho(K,K')>$:

$$i \frac{\partial <\rho(K,K')>}{\partial t} = \int_{-\pi}^{\pi} dq \int_{-\pi}^{\pi} dq' <B(K,K'|q,q')\rho(q,q')> +$$

$$+ (\varepsilon_K - \varepsilon_{K'})<\rho(K,K')> \tag{1.145}$$

Now it is possible to apply the technique presented in considering a two-level problem analogous to the one used to obtain equations (1.138) from relations (1.135). Decoupling correlators in equality (1.145), one derives the following equation for averaged density matrix $\rho(K,K')$:

$$\frac{\partial \rho(K,K')}{\partial t} = -[\Gamma + i(\varepsilon_K - \varepsilon_{K'})]\rho(K,K') +$$

$$+ \frac{1}{2\pi} \int_{-\pi}^{\pi} \Gamma\left(K + \frac{q}{2}, K' + \frac{q}{2}\right)\rho(K + q, k' + q)dq \tag{1.146}$$

where

$$\Gamma = 2 \sum_{m=-\infty}^{\infty} G(m)$$

$$\Gamma(q,q') = 2\left\{\sum_{m=-\infty}^{\infty} G(m)\{\cos[m(q + q')] + \cos[m(q - q')]\} - G(0)\right\}$$

For simplicity we shall only consider the case when $\varepsilon(K) = 2\beta \cos K$ and all $G(m)$ with the exception of $G(0) = 1/2\Gamma$ are equal to zero. In this case Equation (1.146) assumes the form

$$\frac{\partial \rho(K,K')}{\partial t} = -[\Gamma + i(\varepsilon_K - \varepsilon_{K'})]\rho(K,K') + $$

$$+ \frac{\Gamma}{2\pi} \int_{-\pi}^{\pi} \rho(K+q, K'+q)dq \tag{1.147}$$

Let a particle at initial time $t = 0$ be localized in a cell with $n = 0$, i.e.,

$$\rho(K,K')\big|_{t=0} = 1/2\pi \tag{1.148}$$

The study of the particle localization process in the system considered is connected with the solution of Equation (1.147) at initial condition (1.148). Let us introduce new variables u, s and function $F(u,s)$ according to expressions

$$s = \frac{1}{2}(K + K'), \qquad u = \frac{1}{2}(K - K'), \qquad \rho(K',K) = e^{-\Gamma t}F(u,s)$$

Using the new variables for function $F(u,s)$ we obtain the equation

$$\frac{\partial F(u,s,t)}{\partial t} = 4i\beta \sin u \sin sF(u,s) + \frac{\Gamma}{2\pi}\int_{-\pi}^{\pi} F(u,s)ds \tag{1.149}$$

with initial condition

$$F(u,s)\big|_{t=0} = 1/2\pi$$

The solution of Equation (1.149) can be written in the form of the integral equation given below for function $\Phi(u,\tau) = \int_{-\pi}^{\pi} F(u,s,\tau)ds$.

$$\Phi(u,t) = J_0(4\beta t \sin u) + \Gamma \int_0^t J_0\{4\beta \sin[u(t - \tau)]\Phi(u,\tau)d\tau$$

where $J_0(x)$ is Bessel function. Solving this equation applying the transform Laplace method one obtains

$$\Phi(u,t) = \frac{1}{2\pi i}\int_{\sigma-i\infty}^{\sigma+i\infty} \frac{dp \exp(pt)}{[\rho^2 + 16\beta^2 \sin^2 u]^{1/2} - \Gamma}$$

Using this equation, the stated problem is solved in principle since function $F(u,s,t)$ and hence density matrix $\rho(K,K')$ can be determined.

However the expression obtained for $\rho(K,K')$ is too large to be transformed from variables K,K' to point variables m,n for obtaining the density matrix in coordinate representation. So let us calculate the second moment of spatial distribution. Distribution moment $\langle n^{\nu}\rangle$ is given by expression

$$\langle n^{\nu}\rangle = \left(\frac{i}{2}\right)^{\nu} \int_{-\pi}^{\pi} \left\{\frac{\partial^{\nu}\rho(u,s)}{\partial u^{\nu}}\right\}_{u=0} ds$$

The expression for density matrix $\rho(u,s,t)$ is easily expanded as a series in u and integrated with respect to s. So generally any moment can be calculated. However the general expression for ν-th moment is too complicated, therefore we shall write out only the second moment (it is easily shown that if ν is odd we have $\langle n^{\nu}\rangle = 0$):

$$\langle n^2\rangle = \frac{4\tilde{\beta}t}{\Gamma} - \frac{4\tilde{\beta}^2}{\Gamma^2}\left(1 - e^{-\Gamma t}\right)$$

Let us consider limit cases $|\tilde{\beta}| \gg \Gamma$ and $|\tilde{\beta}| \ll \Gamma$. In the first case

$$\langle n^2\rangle = 2(\tilde{\beta}t)^2$$

This relation corresponds to the free motion of a particle compared to (1.93), when the mean square of displacement increases in proportion to the time squared. In case $\Gamma \gg |\tilde{\beta}|$ we have

$$\langle n^2\rangle = 4\tilde{\beta}^2 t\Gamma^{-1} \tag{1.150}$$

Formula (1.150) (compare to (1.92)) describes the normal diffusion of a classical particle with diffusion coefficient $D = 2\tilde{\beta}^2\Gamma^{-1}$.

Thus in the system described the wave packet moves in a purely quantum way at $\tilde{\beta}^2 \gg \Gamma^2$ while at $\tilde{\beta}^2 \ll \Gamma^2$ normal diffusion is observed. Between these two limits there is a transient regime when the motion carries a mixed character.

The same result can be obtained differently. Let us consider the spread of quasi-particles in the crystal lattice when there is random force $f(t)$. The Hamiltonian of the system has the form $(h/2\pi = 1)$

$$\hat{\mathscr{H}}(p) = E(p) + xf(t) \tag{1.151}$$

where

$$E(p) = 2\tilde{\beta}(1 - \cos pd) \quad \tilde{\beta}d^2p^2 \equiv p^2/2m_{\text{eff}}$$

$$m_{\text{eff}} = 1/2\tilde{\beta}d^2 \qquad (d\text{—lattice constant})$$

Quasi-classical equations for time derivatives $\dot{p}$ and $\dot{x}$ of particle moment p and coordinate x accordingly have the form

$$\dot{p} = -\partial \mathscr{H}/\partial x = -f(t)$$

$$\dot{x} = \partial \mathscr{H}/\partial p = 2\tilde{\beta} d \sin pd \tag{1.152}$$

Let us assume that random process $f(t)$ is Gaussian, so

$$\langle f(t)\rangle = 0, \qquad \langle f(t_1)f(t_2)\rangle = 2\mu^2 \delta(t_1 - t_2)$$

From the first equation (1.152) we have

$$p(t) = -\int_0^t f(\tau)d\tau, \qquad p(0) = 0$$

while for $x(t)$ the following expression is easily derived

$$x(t) = x_0 - 2\tilde{\beta} d \int_0^t \sin\left[d\int_0^{\tau'} f(\tau)d\tau \, d\tau'\right] \tag{1.153}$$

Mean square displacement can be obtained from (1.153) with use of equation

$$\left\langle \exp\left[\int_0^t f(\tau)u(\tau)d\tau\right]\right\rangle = \exp\left[\mu^2\int_0^t u^2(\tau_0)d\tau_0\right] \tag{1.154}$$

which can be derived by expanding the exponent as a series.

Applying (1.154) one obtains

$$\langle x^2(t)\rangle = d^2\langle n^2(t)\rangle$$

where $\langle n^2(t)\rangle$ is given by Equation (1.150) with $\Gamma = \mu^2 d^2$. The fact that the quantum result coincides with the quasi-classical one shows the applicability of the latter for large t. The following moments $\langle x^{2K}(t)\rangle$ are also easily obtained, however the calculations will be more cumbersome.

CHAPTER 2

DIFFUSION CONTROLLED PROCESSES WITH INTERMOLECULAR INTERACTION

Chapter 1 considers the main approaches to working out the DCR theory for systems where long-range forces determined by intermolecular interaction can be neglected. The general conclusion is that the rate constant of diffusion-limited reactions is a time function [62]. This function is determined by boundary conditions and, as will be shown in Chapter 3, by the correlation between particles arising in the course of the chemical reaction. If the intermolecular interaction reinforcing the manifestation of correlation effects is taken into account in DCR kinetic theory, the time dependence of the rate constant should become more complicated. Then the intermolecular interaction itself depends on particle concentration in the system, a fact that also complicates the analysis. Only at "infinite" dilution can the potentials be considered known from the intermolecular interaction theory. In the general case, even in the absence of chemical reactions, the appropriate media (for example, solids with a certain amount of defects that can be treated as charged quasi-particles: non-polar liquids containing electrons and ions arising from irradiation, etc.) should be considered analogous to the theory of electrolytes (for example in the Debye-Huckel theory approximation). In these media, screening of carrier electrical charges are manifested and an effective potential is formed which depends on the interaction parameters and the concentration of the corresponding particles. At very low concentrations it can be considered that screening effectively redeterminates the geometric particle size. Then, in the three-dimensional case, instead of the radius of spherical particles [62] trapping radius R should be introduced. Sometimes it appears possible to ignore the potential variation resulting from screening and the chemical reaction, and thus the problem is reduced to describing random walks in certain fields.

2.1. TREATMENT OF INTERMOLECULAR INTERACTION IN THE SMOLUCHOWSKI METHOD

Two approaches are usually used [19,61] for the description of random walks in a potential field. The first is found in the work of Langevin [19].

It is based on analyzing motion equations including a stochastic term besides the term connected with dynamic friction. The deficiency of this approach lies in the fact that in the general case the initial equation without the stochastic term has no solution in the analytical form. This considerably complicates further analysis but the problem is solved numerically [67,68].

The second approach, the one to be considered here, is founded on the application of the Smoluchowski equation [19,69]. When deducing it, it is assumed that the Brownian motion can be described as a Markovian process. Then, as has been shown in §1.8, the distribution function in coordinate space obeys the differential equation usually called the Smoluchowski equation

$$\partial W(\vec{r},t)/\partial t = \operatorname{div} D[\operatorname{grad} W(\vec{r},t) - \beta\vec{F}W(\vec{r},t)] \tag{2.1}$$

where $\vec{r}$ is the radius-vector of particle position; $W(\vec{r},t)d\vec{r}$—the probability of finding the particle in the region $d\vec{r}$ near $\vec{r}$; $\vec{F}$—the force acting upon the particle; D—diffusion coefficient; $\beta = kT$.

This equation is used to study various processes: fluorescence quenching [20,83], segregation of impurities on dislocations [84,85], neutralization of ion pairs [86,87], conductivity in a condensed phase, and various problems of radiation chemistry [87,88], etc.

Analytic solutions of the non-stationary Smoluchowski equation exist only for the Coulomb potential [89,90] and have a complete mathematical structure (double series expansions for the Laplace transformed equation [90] or an expression through Mathieu's function [89]), which complicates the analysis of DCR general kinetics, for example, finding $k_{ex}(t)$. Therefore as a rule either the solution of the stationary equation is used or the solution of the "modified" equation, which provides for excluding the time variable.

Two research directions can be singled out in the study of chemical reaction kinetics of interacting particles using the Smoluchowski equation. One of them originated in the works of Fucks [91] and Debye [92] and is connected with the consideration of concentration gradients in the system. The other, first developed in the work of Onsager [93], is based on analyzing the dynamics of an isolated pair of interacting particles. The difference between the two approaches is rather relative. Practically one and the same problem is being analyzed—only the initial particle distributions are different.

This difference is exemplified by considering the problem of charged particle recombination arising in the condensed phase (for example on exposure to radiation) at a certain initial time moment [86]. It has been discovered that in fluids and solids the general recombination process consists of two stages. In the first stage, the neutralization reaction occurs only between the charged particles of each of the irradiation pairs (between particles—"relatives"). The second stage includes reactions of the charged particles that escaped recombination (with their counterions) during the first phase with any other anti-ions in the system. These stages can be considered as separate processes, which differ only by the initial distribution of active centers. Generally speaking this artificial division of the problem into two parts is determined by the fact that the analytic solutions for arbitrary potentials of the non-stationary Smoluchowski equation are not known. Each of the problems carries certain information about the kinetic characteristics of general DCR recombination. Within the framework of the Smoluchowski-Debye method, the principle information consists of finding the effective constant of reaction rate $k_{ex}(t)$, whereas in the framework of the Onsager-Monchick isolated pair method—in finding the recombination probability or the probability of moving away from particles without reacting ("escape" probability).

2.1.1. The Smoluchowski-Debye Method

For simplicity let us consider a "solution" containing two types of dissolved particles A and B, which react when the distance between their centers is R (usually the sum of the radii of spheric particles). The kinetics of such a reaction should be described by Equation (1.23). Generally speaking the reaction rate is expressed by the component activity. We shall assume the dilution degree of the solution to be such that activity coefficients are about 1*.

We shall use the Smoluchowski equation (2.1) for finding function $k_{ex}(t)$. As in the case of diffusion "recombination" of neutral† particles, we shall take one of the particles, of type A for example, to be located at the origin of the coordinates. The problem consists of finding the particles flux of another type. In the case of spheric symmetry (which is usually assumed) and a potential character of the interaction, when $\vec{F} = -\operatorname{grad} U$, the Smoluchowski equation for probability density $W(r,t)$ can be simplified to:

$$\frac{\partial W(r,t)}{\partial t} = -\operatorname{div} J = \frac{D}{r^2}\frac{\partial}{\partial r}\left[r^2\left(\frac{\partial W}{\partial r} + \beta\frac{\partial U}{\partial r}W\right)\right] \tag{2.2}$$

Here D is the summary (total) diffusion coefficient (hereinafter it is assumed to be constant for simplicity); J is the diffusion flux, U is the potential energy of particle interaction. The diffusion current is assigned by the expression

$$J = -D\left(\frac{\partial W}{\partial r} + \beta\frac{\partial U}{\partial r}W\right) \tag{2.3}$$

Rate constant $k_{ex}(t)$ is determined by the quantity of current $J(R)$ through the absorbing sphere of radius R:

$$k_{ex}(t) = -4\pi R^2 J(R) = 4\pi DR^2\left(\frac{\partial W}{\partial r} + \beta\frac{\partial U}{\partial r}W\right)_{r=R} \tag{2.4}$$

---

*In the general case the activity value is connected with the effective interaction potential [61], and the consideration of the problem becomes greatly complicated.

†The interaction of active centers may have a complex nature, and the particles do not necessarily have a charge. The term neutral particle here means non-interacting at $r > R$ particles.

The given boundary conditions are the same as in the ordinary Smoluchowski method:

$$W(\infty, t) = 1 \tag{2.5}$$

$$W(R, t) = 0 \tag{2.6}$$

Under condition (2.6) the expression for the rate constant takes its "usual" form

$$k_{ex}(t) = 4\pi DR^2 \partial W/\partial r|_{r=R} \tag{2.7}$$

The initial conditions are defined by the specific character of the problem, and in the concentration gradient method it should be assumed that

$$W(r,0) = 1, \quad r > R \tag{2.8}$$

Let us first consider stationary solutions of Equation (2.2). Discussions of non-stationary solutions are usually conducted within the framework of approximated solution methods and also by using model potentials [94,95]. An analysis of this type will be carried out separately.

We shall start by considering the most significant case typical of ionic reactions, when the interaction has a Coulomb form

$$U = q_0/r \tag{2.9}$$

where $q_0 = q_1 q_2/\varepsilon$ is the product of particle charges of each type (ε is the dielectric permittivity of the medium; $q_0 < 0$ for charges of opposite sign).

Equation (2.2) assumes the form

$$\frac{d}{dr}\left[r^2 \frac{dW}{dr} + \eta r_{On} W\right] = 0 \tag{2.10}$$

where the "Onsager radius" is introduced

$$r_{On} = |q_1 q_2|\beta/\varepsilon = |q_0|\beta, \qquad \eta = \mathrm{sgn}(q_1 q_2)$$

Numerical value r_{On} is determined as the distance at which the interparticle interaction energy is equal to the mean energy of the thermal motion of a molecule. At distances larger than r_{On}, the particle ceases to "feel" the potential of the "parent" ion.

The solution of Equation (2.10) under conditions (2.5) and (2.6) is easily found; for example $\eta = 1$:

$$W(r) = [1 - \exp(r_{On}/R)]^{-1}[\exp(r_{On}/r) - \exp(r_{On}/R)] \tag{2.11}$$

and for the rate constant in the stationary case one obtains:

$$k_\infty = 2\pi D r_{On}[1 + \mathrm{cth}(r_{On}/2R)] \tag{2.12}$$

Thus taking into account the interaction potential between particles has led to a new effective size of particles.

Expression (2.12) was used for analyzing experimental data [61,96, 97]—in a number of cases it provided satisfactory agreement with experiment.

It is quite easy to find the solution of Equation (2.2) for potentials of the general type. Using relationship

$$\nabla W + W\beta\nabla U = e^{-\beta U}\nabla(We^{\beta U}) \tag{2.13}$$

the corresponding solution can be presented in the form

$$We^{\beta U} = C\int_r^\infty \frac{e^{\beta U}}{z^2}\,dz + C_1$$

Constants C and C_1 are found from boundary conditions. Taking into account (2.5) and (2.6) one obtains

$$W(r) = e^{-\beta U}[1 - S(r)/S(R)] \tag{2.14}$$

where

$$S(r) = \int_r^\infty e^{\beta U} z^{-2} dz \tag{2.15}$$

In this case rate constant has the form

$$k_\infty = 4\pi D S^{-1}(R) = 4\pi D R_D \tag{2.16}$$

thus value $S(R)$ determines "Debye radius" R_D:

$$R_D = S^{-1}(R)$$

For Coulomb potential (2.9) we obtain the known result:

$$S^{-1}(R) = \frac{1}{2}\, r_{On}[1 + \mathrm{cth}(r_{On}/2R)] \tag{2.17}$$

Debye radius R_D performs the role of the effective particle radius in the case of arbitrary potentials. Evidently R_D signifies the distance starting from which mobile particles are exposed to the influence of the selected particle.

The fact that the Smoluchowski equation has stationary solutions permits discussing somewhat differently the most interesting question in DCR kinetics—the role of particle correlation. When stationary particle distribution is formed in the system, the correlation between active centers determines the type of effective self-consistent potential. As has already been noted, this potential should be determined through a self-consistent procedure using, for example, the adiabatic equation. In the absence of a chemical reaction this procedure is for example given by the Debye-Hückel method [98]. The course of a chemical reaction may considerably complicate the derivation of the effective potential even within the limits of the Debye-Hückel approximation. From this point of view, the applicability of the Smoluchowski-Debye method [99] requires additional discussion. Let us consider several propositions of [99] which provide for a more detailed presentation of the connection of the effective potential with concentration and the component reactivity.

As before, we shall take the solution of two types of ions and write the Smoluchowski equation for ion concentrations of an arbitrary sort:

$$\frac{d}{dr}[r^2 J_i(r)] = 0 \qquad (i = A, B) \tag{2.18}$$

where $J_i(r)$ is determined as in (2.3):

$$J_i(r) = -D\left[\frac{dc_i}{dr} + q_i\beta \frac{dU_e}{dr} c_i(r)\right] \tag{2.19}$$

[$c_i(r)$ is the concentration of ions of i types; $U_e(r)$ is the electrostatic potential; q is the charge of ions ($q_A = -q_B$)].

Equation (2.18) can be treated as a continuity equation for $J_i(r)$ in the stationary state. The Smoluchowski-Debye method provides only for analyzing a flux of one type of particle, for example $J_B(r)$ into ions of another type A. The flux of A ions is not considered at that (it is assumed to equal zero). Integration of the system of Equations (2.18) gives the expressions:

$$\frac{dc_i(r)}{dr} + q_i\beta \frac{dU_e}{dr} c_i(r) = \frac{b_i}{r^2} \qquad (i = A, B) \tag{2.20}$$

with $b_B = b \neq 0$, but $b_A = 0$.

The system of Equations (2.20) should be complemented as in the Debye-Hückel theory by the adiabatic equation for potential U_e:

$$-\frac{1}{r^2}\frac{d}{dr}\left(r^2 \frac{dU_e}{dr}\right) = \frac{4\pi|q|}{\varepsilon}(c_A - c_B) \tag{2.21}$$

We shall determine boundary conditions by the relationships

$$c_B(R) = 0$$

$$c_B(\infty) = c_A(\infty) = c_0$$

$$dU_e/dr|_{r=R} = q_A/(\varepsilon R^2) \tag{2.22}$$

where R is the sum of the radii of ions, $R = R_A + R_B$.

The solution of (2.20) for ions of type A gives the Boltzmann distribution:

$$c_A(r) = c_0 \exp(-\beta q_A U_e) \tag{2.23}$$

for ions of type B one obtains:

$$c_B(r) = \exp(-\beta q_B U_e) \left[c_0 - b \int_r^{\infty} \exp(\beta q_B U_e) r^{-2}\, dr \right] \tag{2.24}$$

The equation for the effective potential is deduced by substituting (2.23) and (2.24) into (2.21). Let us write it having introduced non-dimensional quantities

$$\varphi = q\beta U_e, \qquad x = r/R, \qquad q = |q_i|, \qquad \lambda^2 = \varepsilon(8\pi q^2 c_0 \beta)^{-1}$$

Then we get

$$\frac{\lambda^2}{R^2} \frac{1}{x^2} \frac{d}{dx} \left(x^2 \frac{d\varphi}{dx} \right) - \mathrm{Sh}\varphi = \frac{b}{2c_0 R} \int_x^{\infty} \frac{d\eta}{\eta^2} \exp[\varphi(\eta) - \varphi(x)] \tag{2.25}$$

having boundary conditions:

$$\varphi(\infty) = 0, \qquad d\varphi/dx|_{x=1} = q^2\beta/(\varepsilon R^2) \tag{2.26}$$

We have obtained a non-linear integral differential equation of second order, its right part describing the perturbation of the potential due to the course of the chemical reaction. If besides the reacting ions of two types, the system has an inert electrolyte containing several types of ions, parameter $2c_0$ of the right part denominator of (2.25) should be replaced by general concentration n of all ions held in the system

$$n = \sum_j [c_j^+(\infty) + c_j^-(\infty)]$$

where $c_j^+(\infty)$ and $c_j^-(\infty)$ are stationary concentrations of positive and negative ions of the j-electrolyte component respectively.

To satisfy the condition of electroneutrality, it should be assumed that

$$\sum_j c_j^+ = \sum_j c_j^-; \quad \text{if} \quad j = 1, \quad c^+ = c^- = c_0, \quad n = 2c_0$$

In the general case, the problem consists of a self-consistent derivation of the potential determining the rate constant of the process (2.4).

Let us now analyze the behavior of parameter b. Taking into account (2.22) it follows from (2.24) that

$$bR^{-1}c_0^{-1} \int_1^\infty \exp[\varphi(x,b)] \frac{dx}{x^2} = 1 \tag{2.27}$$

Since effective potential $\varphi(x,b)$ depends on parameter, the solution of system (2.25) and (2.27) in the general case should be sought at boundary conditions (2.26) [for "neutral" systems when $U_e(r) = 0$, it is evident that $b = c_0 R$].

It can be expected that for weak fields ($|\varphi| < 1$) value $bR^{-1}c_0^{-1}$ will be close to 1. In this case it is possible to find the condition under which the potential perturbation is small due to the course of the chemical reaction. Obviously this occurs in the case of a limit dilution ($\lambda \to \infty$) which preserves the Coulomb potential.

A special case arises if an inert electrolyte is also present in the system with $bR^{-1}n^{-1} \sim c_B(\infty)n^{-1}$. In the case of a large electrolyte excess when $c_B n^{-1} \ll 1$, Equation (2.25) can be solved ignoring the right part which leads to the known [80] expression for the effective potential:

$$\varphi = \frac{q^2\beta}{\varepsilon r} \frac{\exp[-x(r - R)]}{1 + xR}, \qquad x^2 = \lambda^{-2} = \varepsilon^{-1} \cdot 4\pi q^2 n\beta$$

In the general case the effective potential is a complex concentration function.

Two aspects are of special importance for DCR kinetics. The stationary rate constant defined by expression (2.16) is proportional to R_D, i.e., carries in itself information on the effective potential, and hence on the degree of interparticle correlation.

When analyzing non-stationary stages of DCR, effective rate constant $k_{ex}(t)$ will be a complex function of time. The concentration of reacting active centers in DCR may vary by several orders of magnitude and thus in the course of the reaction the effective potential can be expected to change considerably. Only at long time ($t \to \infty$) when the concentration of reacting particles is low is the potential determined by the nature of intermolecular interaction forces.

2.1.2. The Isolated Pair Method

The isolated pair method provides for analyzing DCR kinetics at times when the existence of strong correlation in the initial particle position is appreciably manifested and potential perturbation due to the course of the course of the chemical reaction is small. Quite often we have a case of an initial distribution when particles are being created in pairs, i.e., the distance between a pair of particles (of reacting types) is much smaller than the mean distance between pairs. Then the probability of a reaction between particles in a pair is far greater than the probability of a reaction between particles of different pairs. In the case of such a specific initial distribution, DCR kinetics may differ considerably from the reaction kinetics with a uniform distribution of components up to long times, when the number of particles remaining in the system is much smaller than the initial number. Therefore traditionally these reactions fall into a special class of "correlated" diffusion controlled reactions which are investigated on the basis of analyzing the dynamics of a particle pair.

Mathematically the problem is reduced to solving the Smoluchowski equation but at a different (as compared to the Smoluchowski-Debye method) initial condition and the condition "at infinity". Then of special interest is not the stationary rate constant but the calculation of the survival probability P_e of a particle pair, or the probability P_R of particle recombination ($P_R = 1 - P_e$). A general discussion of the method related to the work of Onsager is given in review [87] based on the example of an analysis of ion pair dynamics (Coulomb interaction).

Let us consider the application of the method following Monchick's analysis [24] in which the potential is given in the general form and use is made of a boundary condition of the third type ("radiation" boundary condition) on the reaction surface.

Let us consider a three-dimensional spherically symmetric problem using Equation (2.2) with initial condition

$$W(r,0) = \delta(r - r_0)/(4\pi r^2) \tag{2.28}$$

(at the initial time moment a pair of particles is at distance r_0 from each other) and the boundary condition at infinity

$$W(r,t) \to 0 \quad \text{at} \quad r \to \infty$$

Let us assign the condition on the reaction surface

$$hW(R,t) = D\left(\left.\frac{\partial W}{\partial r}\right|_{r=R} + \beta W \left.\frac{dU}{dr}\right|_{r=R}\right)$$

where $h = k/(4\pi R^2)$ (k is the real rate constant of the reaction).

We will find reaction probability P_R, defining it by the relationship

$$P_R = 1 - P_e = \int_0^\infty \tilde{R}(t)\,dt, \qquad P_e = \lim_{t\to\infty} \omega(t)$$

$$\tilde{R}(t) = -\frac{d\omega(t)}{dt}, \qquad \omega(t) = \int_R^\infty 4\pi r^2 W(r,t)\,dr \tag{2.29}$$

where $\tilde{R}(t)$ is the reaction probability between particles by time t; $\omega(t)$ is the probability of the particle survival by time t; P_e is the full survival probability of a pair of particles.

Equation (2.2) can be rewritten in the form

$$\frac{\partial W(r,t)}{\partial t} = L(W), \qquad L(W) = \nabla(D\nabla W + \beta D\nabla UW) \tag{2.30}$$

convenient for further discussion.

Let us introduce functions

$$\alpha(r,t) = W(r,t)\exp\beta U, \qquad \bar{\alpha}(r) = \int_0^\infty \alpha(r,t)\,dt \tag{2.31}$$

assuming $\alpha(r,\infty) = 0$.

Substitution of (2.31) into (2.30) yields the equation

$$\partial\alpha/\partial t = M(\alpha) \tag{2.32}$$

where

$$M(\alpha) = D\left[\frac{\partial^2\alpha}{\partial r^2} + \frac{\partial\alpha}{\partial r}\left(\frac{2}{r} - \beta\frac{dU}{dr}\right)\right] \equiv D\nabla^2\alpha - \beta\nabla UD\nabla\alpha \tag{2.33}$$

Integration of (2.32) with respect to time gives the "modified" Smoluchowski equation:

$$D^{-1}M(\bar{\alpha}) = \frac{d^2\bar{\alpha}}{dr^2} + \frac{d\bar{\alpha}}{dr}\left(\frac{2}{r} - \beta\frac{dU}{dr}\right) = -\frac{\delta(r - r_0)}{4\pi Dr^2}e^{\beta U} \tag{2.34}$$

where boundary conditions can be written in the form

$$\bar{\alpha}(\infty) = 0, \qquad \left.\frac{d\bar{\alpha}}{dr}\right|_{r=R} = h_1\bar{\alpha}(R), \qquad h_1 = \frac{h}{D} \tag{2.35}$$

Function $\bar{\alpha}$ is Green's function of the corresponding (2.34) uniform equation. It is derived in the standard way on the basis of solutions $\bar{\alpha}_+(r)$ (in region $r > r_0$) and $\bar{\alpha}_-(r)$ (in region $r < r_0$), which are matched at $r = r_0$, i.e., conditions

$$\bar{\alpha}_+(r_0) = \bar{\alpha}_-(r_0)$$

$$\left.\frac{d\bar{\alpha}_+}{dr}\right|_{r=r_0} - \left.\frac{d\bar{\alpha}_-}{dr}\right|_{r=r_0} = e^{\beta U}(4\pi Dr_0)^{-1} \tag{2.36}$$

are fulfilled.

Solutions $\bar{\alpha}_+$ and $\bar{\alpha}_-$ are readily expressed with the aid of function $S(r)$ determined in (2.15):

$$\bar{\alpha}_- = A[S(R) - S(r)] + B$$

$$\bar{\alpha}_+ = CS(r)$$

and coefficients A, B and C are found using matching conditions (2.36) and conditions (2.35).

Our interest lies not with Green's function $\bar{\alpha}(r)$ itself, but with probability P_R, which can be found if flux $J(t)$ per time unit into the absorbing sphere is known

$$J(t) = 4\pi DR^2 \left(\frac{\partial W}{\partial r} + \beta W \frac{dU}{dr}\right)_{r=R} = 4\pi R^2 h\alpha(R,t)e^{-\beta U}$$

Integrating with respect to variable t on the basis of (2.29) one obtains:

$$P_R = 4\pi R^2 hB \exp\left[-\beta U(R)\right]$$

Let us write the explicit expression for P_R:

$$P_R = S(r_0)\left\{S(R)\left[1 + \frac{D \exp \beta U(R)}{hS(R)R^2}\right]\right\}^{-1} \tag{2.37}$$

First we shall analyze the case of purely diffusion control of the reaction assuming $h \to \infty$.

In this case, when $P_R|_{h\to\infty} = P_{RD}$, one obtains $P_{RD} = S(r_0)S^{-1}(R)$.

It can be clearly seen that at $U = 0$, $S(r) = r^{-1}$ and we obtain the known result $P_{RD}(U = 0) = Rr_0^{-1}$.

In case of the Coulomb potential (2.9), function $S(R)$ is given by expression (2.17), and then

$$P_{RD} = [1 - \exp(-r_{On}/r_0)][1 - \exp(-r_{On}/R)]^{-1}; \tag{2.38}$$

at $r_{On} \gg R$ we obtain the Onsager result:

$$P_{RD} = 1 - \exp(-r_{On}/r_0)$$

For particles with initial distances $R < r_0 < r_{On}$ the absorption probability is close to 1, whereas at initial distances $r_0 > r_{On}$ it follows: $P_{RD} \simeq \simeq r_{On}r_0^{-1}$. This means that even at large initial distances, the probability of a reaction is larger than in the case of "neutral" particles.

Now using (2.29) let us calculate survival probability of a particle pair

$$P_e = 1 - P_R = 1 - S(r_0)/S(R)$$

and compare it with expression (2.14) which determines the survival probability of a pair of particles according to the Smoluchowski-Debye method. It can be seen that

$$W(r_0, \infty) = P_e \exp[-\beta U(r_0)] \tag{2.39}$$

This relationship illustrates how different initial conditions in the Smoluchowski-Debye and Onsager-Monchick equations form different survival probabilities.

Now let us once more turn to a discussion of general expression (2.37) for reaction probability $P_R(r_0)$.

Taking into account that $h = k/(4\pi R^2)$ and using (2.16) we shall present (2.37) in the form

$$[4\pi D S^{-1}(r_0) P_R(r_0)]^{-1} = k_\infty^{-1} + k_c^{-1} \tag{2.40}$$

where $k_c = k \exp[-\beta U(R)]$, $k_\infty = 4\pi D S^{-1}(R)$.

Thus we obtain the law of kinetic resistance composition [86]. In case $U = 0$ one obtains

$$k_{ex}^{-1} = k_D^{-1} + k^{-1}$$

i.e., relationship (1.22), if $k_{ex} = 4\pi D r_0 P_R(r_0)$.

The meaning of the new representation for the general rate constant is absolutely clear: multiplier $k_1(r_0) = 4\pi D r_0$ gives the rate constant (the probability of a collision per time unit) for a purely diffusion control of the reaction between particles with trapping radius r_0 (i.e., under the condition that the probability of a reaction upon collision is 1). Value $k_1(r_0)$ multiplied by reaction probability $P_R(r_0)$ for initial r_0 of the distance between reagents gives the absolute rate constant of the entire bimolecular DCR (in the absence of the interaction potential). Relationship (2.40) then determines the general expression for rate constant

$$k_{ex}[U(R)] = 4\pi D S^{-1}(r_0) P_R(r_0) \tag{2.41}$$

It should be noted that in the presence of interaction $U(r)$ the expression for the rate constant of the chemical transformation itself is also modified:

$$k_c = k(0) \exp[-\beta U(R)], \qquad k(0) = k(U)\big|_{U=0} \equiv k$$

The application of the method for analyzing experimental data is treated in detail in reviews [87–101].

2.1.3. Non-Stationary Solutions of the Smoluchowski Equation

A discussion of the problem is considerably complicated due to the absence of analytic solutions of the Smoluchowski equation for arbitrary interaction potentials. Solutions in the general form [89,90] have been obtained only for Coulomb potential, however the complexity of the analytic expressions derived hinders their application. Hence, non-stationary DCR kinetics is usually discussed on the basis of the approximate solutions of the Smoluchowski equation [84,100–105].

Of special interest are "short" time solutions [85] (especially for solid phase reactions when experimental results for the transformation degree are described by power dependences $t^{\pm\nu}$, $\nu < 1$) and the asymptotic behavior of solutions at large time. The analysis of "large" time is important since under investigation is the kinetics of "deep" transformation stages, when the potential in the interacting particle system becomes practically non-screened. Usually of interest are the first corrections to the stationary value of the rate constant (since the stationary solutions are known). Solutions at large times are as a rule obtained with the aid of different procedures of "sewing together" [102,103] of approximate solutions (obtained in regions of large potential values and in far regions when the interaction can be ignored). It is often assumed that the rate constant has the following form:

$$k_{ex}(t) = 4\pi DR_e(1 + Re/\sqrt{\pi Dt}) \tag{2.42}$$

i.e., the first correction to the rate constant appears to be proportional to $t^{-1/2}$ and the effective radius R_e is defined by the behavior of the potential and the character of the approximation.

Let us consider approximate solutions of the Smoluchowski equation using for simplicity the three-dimensional equation in a system with spheric symmetry and an attractive potential.

The physical base for singling out regions of short and large times is connected with the concept of the role of the interaction potential in reaction kinetics. If the effective range of interparticle forces is characterized by Debye radius R_D (2.16) then "short" times are determined by condition $t \leqslant R_D{}^2D^{-1} = \tau_D$. DCR kinetics at these times should be determined by reactions between relatively closely located particles, the mobility of which as compared to the purely diffusional one is considerably increased due to attraction forces. This fact can be illustrated in two somewhat different ways. Authors of [85] used the approximation of a "strong" drift flux supposing that in general expression (2.3)

$$J = -D\left(\frac{\partial W}{\partial r} + \beta \frac{dU}{dr} W\right) \simeq -D\beta \frac{dU}{dr} W \tag{2.43}$$

the first term describing the purely diffusion flux can be neglected. Analyzing impurity migrations to dislocations on the basis of their assumption leads to a power dependence of the quantity of absorbed defects $N(t)$ on time $N \sim t^{2/3}$ for the Coulomb potential. A detailed discussion of this approximation describing the kinetics of the process at short times (at low transformation degrees) is given in [85] and monograph [84].

Paper [106] suggests a method of describing non-stationary chemical transformation kinetics for systems of interacting particles in the limit of small diffusion coefficients, when gradients arising in the course of transformation concentration cannot be smoothed due to diffusion and the principal contribution into the resulting flux is made by the drift term. The latter provides an appreciable simplification of the problem proceeding not from the Smoluchowski equation, but from the Langevin equation (see Section 1.8) in the limit $D \to 0$. Unlike paper [85], paper [106] considers relatively large times, when $t \gg \beta_0^{-1}$ where β_0^{-1} is the effective mobility of particles [for spheric particles of mass m and radius R in accordance with Stokes law $\beta_0 = kT/mD = 6\pi\eta R/m$].

Let us consider the most interesting case of attraction between reacting particles using the example of bimolecular irreversible reactions $A + B \to$ Products, when the transformation rate can be formally presented in the form

$$dc_A/dt = dc_B/dt = -k_{ex} c_A c_B \tag{2.44}$$

where k_{ex} is the "observed" rate constant which, generally speaking, is time dependent.

Besides condition $t \gg \beta_0^{-1}$ we shall assume $t < R_D^2/D$, i.e., we shall confine ourselves to analyzing the kinetics at times when only pairs of A and B particles initially located close enough (at a distance smaller than R_D) from each other can enter into the reaction. In this case for the stationary field of approach velocities of particles, it follows from the Langevin equation (1.78) that

$$dr/dt = -m^{-1}\beta_0^{-1} dU/dr \tag{2.45}$$

where r is the distance between approximating particles.

In the given approximation ($D \to 0$) r and dU/dr in Equation (2.45) can be understood as their most probable values. It should also be noted here that the integration of the Smoluchowski equation (2.2) using approximation $D \to 0$ leads to the relationship

$$\frac{d\bar{r}}{dt} = -\frac{D}{kT}\left(\overline{\frac{dU}{dr}}\right) \tag{2.46}$$

where

$$\bar{r} = \int_R^\infty rW(r,t)dV, \quad \left(\overline{\frac{dU}{dr}}\right) = \int_R^\infty \frac{dU}{dr} W dV \tag{2.47}$$

Relationships of type (2.45) and (2.46) can be treated as the generalization of the Stokes-Einstein relationship [107] using [101,107] for evaluating initial distances and the ion pair lifetime.

We shall confine ourselves to the power-like approximation

$$U = -\varepsilon_0 (r_0/r)^n$$

for the potential energy of the interaction between A and B particles. According to (2.45) the distance between the centers of particles A and B having initial separation r_x diminishes as

$$r^{n+2} - r_x^{n+2} = -\omega t, \qquad \omega = [(n+2)/(n\beta_0 m)]\varepsilon_0 r_0^{\ n} \tag{2.48}$$

By time t the distance between reacting particles will be equal to R (reaction radius) if the initial distance between them does not exceed r_x:

$$r_x = (\omega t + R^{n+2})^{1/(n+2)}$$

Accordingly, probability $P(t)$ by time t R will be the distance between the centers of particles A and B and is determined by the average number of particles B initially located in the sphere layer bounded by concentric spheres with centers coinciding with the center of particle A. The radius of inner sphere is equal to R, the outer one—to r_x. Let us consider a model case when particles of reagent A are generated from some inert with respect to reaction (2.47) particles A^0, the interaction of which with B particles is considerably smaller than the corresponding interaction of active particles A. The initial distribution of reagent B relative to particle A can be considered uniform. Then the probability that by time t some A particle will appear in immediate proximity to particle B will be defined by the product of the concentration of B particles by the volume of the sphere layer with inner radius R and outer radius r_x:

$$P(t) = \frac{4\pi}{3}\,[(\omega t + R^{n+2})^{3/(n+2)} - R^3]c_B \tag{2.49}$$

In case of a sufficiently strong interaction $|U(R)| \gg kT$ the initial quantity of the concentration of B particles located in immediate proximity to particles A exceeds appreciably the mean density of B particles. Accordingly, the reaction rate diminishes rapidly at the initial stage, later it is described by a slowly changing function of time:

$$P(t) = c_B \int_R^{r_x} \exp\,[U/(kT)]r^2 dr$$

At sufficiently large time when $t \gg R^{n+2}/\omega$ value R^{n+2} can be neglected in expression (2.49). Then

$$P(t) = \frac{4\pi}{3}\,(\omega t)^{3/(n+2)} c_B$$

If the kinetic rate constant is sufficiently large, it can be considered that each encounter of particles A and B results in an elementary act of a reaction. The observed rate constant $k_{ex}(t)$ presents the probability that some particle A remaining unreacted by time t will enter into reaction by time $t + dt$, referred to interval dt and unit concentration B, i.e., is defined by the derivative of function $P(t)/c_B$:

$$k_{ex}(t) = [4\pi/(n+2)]\omega^{3/(n+2)} t^{-(n-1)/(n+2)} \tag{2.50}$$

Substituting (2.50) into (2.44) one obtains the desired expression for the kinetics of the process. For reactions of pseudo-first order, when the concentration of one of the components is much higher than the concentration of the other component, i.e., $c_B \gg c_A^0$ (where c_A^0 is the value of the initial concentration of A particle), we derive the expression

$$c_A(t) = c_A^0 \exp(-amt^b)$$

where

$$m = c_B, \quad a = \frac{4\pi}{3}\left(\frac{n+2}{n}\,\frac{\varepsilon_0 D r_0^n}{kT}\right)^{3/(n+2)}, \quad b = \frac{3}{n+2} \tag{2.51}$$

For bimolecular processes in the case of equal concentrations of particles A and B, it follows from (2.47) that

$$c_A^0 / c_A(t) = 1 + amt^b$$

where a and b are determined by (2.51); $m = c_A^0 = c_B^0$.

Expression (2.49) for $P(t)$ is obtained without taking into account diffusion equilibration of the concentration gradients arising in the reacting system. The given approximation is valid for ion pairs of reagents A and B, the initial distance between which is smaller than R_{cr}. In this case the potential energy of the interaction between A and B is equal to the average kinetic one:

$$|U(R_{cr})| = (r_0/R_{cr})^n \varepsilon_0 = kT$$

Parameter $\alpha = 4/3\ \pi R_{cr}^3 m$, equal to the probability of detecting a reagent particle in a sphere of radius R_{cr}, determines the part of reagents having reacted in accordance with (2.50) or (2.51) and it can be presented by the expression

$$\alpha = \frac{4\pi}{3}\,[\varepsilon_0/(kT)]^{3/n} r_0^3 m$$

The difference in the character of the interaction of reacting particles determines the difference in the chemical transformation kinetics. For example, in the interaction between an anion and a cation ($n = 1$)

$$k_{ex}(t) = 4\pi z_A z_B e^2 D_{AB} / (\varepsilon KT)$$

which coincides with $k_{ex} = 4\pi R_D D$, obtained when solving the stationary Smoluchowski equation.

In the interaction of ions with neutral particles ($n = 4$)

$$k_{ex}(t) = \frac{\pi}{2\sqrt{3}}\left(\frac{\sqrt{\alpha_B z_A e}}{\varepsilon}\right)^3 t^{-1/2}$$

where $z_A e$ is the charge of the ion, ε is the dielectric permittivity of the medium.

For neutral polar particles with power-like interaction, (at $m = 6$) we get

$$k_{ex}(t) = (6kT)^{-3/8}[(\alpha_A \mu_B^2 + \alpha_B \mu_A^2)D]^{3/8}(8t)^{-5/8} \tag{2.52}$$

where $\mu_{A,B}$ are dipole moments of reagent particle; $\alpha_{A,B}$ is their polarizability.

Now let us consider a relatively simple method of an approximate solution for Smoluchowski equation (2.2) suggested in [103].

We shall use the simplest boundary condition (2.5), Coulomb potential (2.9) and employ the substitution

$$W(r, t) - W(r, 0) = y(r, t) \exp(\beta U/2) \tag{2.53}$$

Applying the Laplace transform and using (2.2) one obtains the equation

$$\frac{1}{r}\frac{d^2[r\bar{y}(r,s)]}{dr^2} - \left(\frac{s}{D} + \frac{r_{On}^2}{4r^4}\right)\bar{y}(r,s) = 0 \tag{2.54}$$

where

$$\bar{y}(r,s) = \int_0^\infty y(r,t)e^{-st}dt, \qquad r_{On} = \beta q_0$$

Substitution (2.53) is most suitable for a random (2.8) or Boltzmann (2.39) initial distribution (any other initial distribution makes (2.54) heterogeneous).

Equation (2.54) can be simplified in two regions defined using value $r_s = [r_{On}^2 D/(4s)]^{1/4}$. Let us determine the "inner" region $r < r_s$, in which the equation is valid

$$\frac{1}{r}\frac{d^2[r\bar{y}_1(r,s)]}{dr^2} - \frac{r_{On}^2}{4r^4}\bar{y}_1(r,s) = 0 \tag{2.55}$$

and the "outer" region $r > r_s$, where

$$\frac{1}{r}\frac{d^2[r\bar{y}_2(r,s)]}{dr^2} - \frac{s}{D}\bar{y}_2(r,s) = 0 \tag{2.56}$$

Equation (2.55) is stationary Equation (2.2), Equation (2.56) is Equation (2.2) at $U(r) = 0$.

We shall look for the approximate solution of Equation (2.54) matching the solutions of Equations (2.55) and (2.56) at $r = r_s$, i.e., satisfying the conditions

$$\bar{y}_1(r_s) = \bar{y}_2(r_s)$$

$$\partial \bar{y}_1/\partial r\Big|_{r=r_s} = \partial \bar{y}_2/\partial r\Big|_{r=r_s} \tag{2.57}$$

The general solution of Equation (2.55) has the form

$$\bar{y}_1(r,s) = A_1 \exp(r_{On}/2r) + B_1 \exp(-r_{On}/2r), \qquad R \leqslant r \leqslant r_s \tag{2.58}$$

and of Equation (2.56):

$$\bar{y}_2(r,s) = \frac{1}{r}\,[A_2 \exp(r\sqrt{s/D}) + B_2 \exp(-r\sqrt{s/D})] = B_2 \exp(-r\sqrt{s/D})/r$$

since $y(r,t) \to 0$ at $r \to \infty$.

Let us calculate rate constant $k_{ex}(t)$. The expression for $k_{ex}(t)$ depends on initial conditions and boundary condition at $r = R$.

Using a boundary condition of the third kind

$$kW(R,t) = 4\pi DR^2 \partial W/\partial r\Big|_{r=R} \tag{2.59}$$

rate constant $k_{ex}(t)$ can be determined from the expression $k_{ex}(t) = kW(R,t)$.

The representation of $k_{ex}(t)$ for a Laplace transform takes the form:

$$k_{ex}(s) = k \left[\frac{W(R,0)}{s} + \exp(-r_{On}/2R)\bar{y}(R,s)\right] \tag{2.60}$$

From (2.60) follows the difference of $k_{ex}(s)$ at a random initial distribution

$$W(r,0) = 1 \tag{2.61}$$

or at a Boltzmann one

$$W(r,0) = \exp(-r_{On}/r) \tag{2.62}$$

To find constant $k_{ex}(s)$, it is necessary to determine coefficients A_1 and B_1 in (2.58). Coefficients A_1, B_1, and B_2 are easily obtained using conditions (2.5), (2.57) and (2.58).

Let us write the expression for $\bar{y}_1(R,s)$ at random initial distribution (2.61):

$$\bar{y}_1(R,s) = \left[\left(1 - r_{On}\sqrt{\frac{s}{D}}\right)\exp\left(\frac{r_{On}}{R}\right) - k(4\pi D r_{On} + k)^{-1}\right]^{-1} \times$$
$$\times \left[\exp\left(-\frac{r_{On}}{2R}\right) - \left(1 - r_{On}\sqrt{\frac{s}{D}}\right)\exp\left(\frac{r_{On}}{2R}\right)\right]\exp\left(\frac{r_{On}}{R}\right)s^{-1} \tag{2.63}$$

Because of limitation $r > R$, this expression is non-informative at $s > Dr_{On}^2/(4R^4)$. We are interested in the asymptotic behavior of the rate constant of the process at large times, which requires finding the limit expression $\bar{y}_1(R,s)$ at $s \to 0$. At $s < Dr_{On}^2/(4R^4)$ from (2.60) and (2.63) one can obtain:

$$k_{ex}(s) = 4\pi r_{On} Ds^{-1}\left[(4\pi r_{On} Dk^{-1} + 1)\exp\left(\frac{r_{On}}{R}\right) - 1\right]^{-1} \times$$
$$\times \left\{1 + r_{On}\sqrt{\frac{s}{D}}\left[1 - \exp\left(\frac{r_{On}}{R}\right)\left(1 + 4\pi r_{On} Dk^{-1}\right)\right]^{-1}\right\} \tag{2.64}$$

On the basis of (2.64) using the reverse Laplace transform we get

$$k_{ex}(t) \geqslant 4\pi R_e D\left[1 + \frac{R_e}{\sqrt{\pi Dt}}\exp\left(\frac{r_{On}}{R}\right)\left(1 + 4\pi r_{On} Dk^{-1}\right)\right] \tag{2.65}$$

where effective radius R_e is introduced:

$$R_e = r_{On}\left[(1 + 4\pi r_{On} Dk^{-1})\exp\left(\frac{r_{On}}{R}\right) - 1\right]^{-1}$$

In the case of Boltzmann initial distribution (2.62) the expression for $k_{ex}(t)$ coincides with (2.42). It follows from (2.65) the same as from (2.42) that $\Delta k_{ex} = k_{ex}(t) - k_{ex}(\infty) \sim t^{-1/2}$.

To check the result, the authors of [103] presented the results of the numerical solution* of the Smoluchowski equation at random initial particle distribution. The value of r_{On} varied from 0.7 nm (for water) to $\sim$30 nm (for hydrocarbons), value R was taken to be 0.5 nm and for simplicity it was assumed in (2.63) that $k \to \infty$ [this corresponds to using the boundary condition of Smoluchowski (2.6)].

In discussing the results, paper [103] gives an evaluation of time interval τ starting from which the Smoluchowski-Debye equation can be applied. For this purpose, the results of paper [66] have been used

---

*A discussion of the procedures and results of the numerical solution of the Smoluchowski equation can be found in [108–111].

where the more general hyperbolic telegrapher's equation was being investigated. This equation differs from the Smoluchowski-Debye equation by the additional term $\partial^2 W/\partial t^2$. Paper [66] shows that the telegrapher's equation is identical to the Smoluchowski-Debye equation at time $t > \tau$, where $\tau = mD(kT)^{-1}$ (m is the mass of the ion).

At values $D = 10^{-8} m^2/s$, $T = 300$ K, $m = 5 \cdot 10^{-26}$ kg one obtains $\tau = 0.7$ ps. Hence [102] assumes that the numerical solution obtained is applicable when $t > 1$ ps.

An analysis of the numerical solution shows that within the limits of physical applicability of the Smoluchowski-Debye equation, the Equation (2.65) agrees to a large extent with the result of the numerical solution at $t > 10$ ps and at $|r_{On}| = 0.5$ nm. Since $R = 0.5$ nm, values $|r_{On}| \simeq 0.7$ are characteristic for "weak" potentials when $|r_{On}| \simeq R$.

Paper [103] also considers the situation for "strong" potentials, when $|r_{On}| \gtrsim 20$ nm, i.e. $|r_{On}|R^{-1} \geqslant 40$ (hydrocarbon solvents). In this case (the reagents are oppositely charged and randomly distributed in space) rate constant k_{ex} decreases rapidly to its stationary value and the approximate solution (2.65) diverges appreciably from the results of the numerical calculation. Thus the case of strong potentials requires special analysis.

Let us analyze this case using model potential

$$\Phi_{eff} = \beta U_{eff} = -2 \ln\left(1 + \frac{x_0}{r}\right) \tag{2.66}$$

where x_0 is the parameter proportional to the Onsager radius (a detailed discussion of this potential is given in Section 2.4).

We shall consider Equation (2.2) with potential (2.66) and introduce a new function

$$\alpha(r,t) = W(r,t)\exp(\Phi_{eff}) \tag{2.67}$$

Function $\alpha(r,t)$ satisfies the equation

$$\frac{\partial \alpha(r,\tau)}{\partial \tau} = \frac{\partial^2 \alpha}{\partial r^2} + \frac{\partial \alpha}{\partial r}\,\frac{2}{r + x_0}, \qquad \tau = Dt \tag{2.68}$$

which we shall solve under the simplest boundary conditions following from (2.5) and (2.6):

$$\alpha(R,\tau) = 0, \qquad \alpha(\infty,\tau) = 1 \tag{2.69}$$

Two types of initial conditions are being analyzed

$$\alpha(r,0) = 1 \tag{2.70}$$

$$\alpha(r,0) = \exp(\Phi_{eff}) = r^2/(r + x_0)^2 \tag{2.71}$$

which correspond to the Boltzmann distribution and to the random initial distribution respectively. The stationary solution of Equation (2.68) is readily obtained under conditions (2.69):

$$\alpha_{st}(r) = 1 - (R + x_0)/(r + x_0) \tag{2.72}$$

We shall start a discussion of the nonstationary solutions with the case of Boltzmann initial distribution (2.70).

Let us introduce function

$$\alpha_1(r,\tau) = \alpha(r,\tau) - 1$$

and apply Laplace transform

$$\bar{\alpha}_1(r,s) = \int_0^\infty e^{-s\tau}\alpha_1(r,\tau)\,d\tau$$

Function $\bar{\alpha}_1(r,s)$ satisfies equation

$$\frac{\partial^2\bar{\alpha}_1}{\partial r^2} + \frac{2}{r + x_0}\frac{\partial\bar{\alpha}_1}{\partial r} - s\bar{\alpha}_1 = 0 \tag{2.73}$$

and boundary conditions

$$\bar{\alpha}_1(R,s) = -1/s, \qquad \bar{\alpha}_1(\infty,s) = 0$$

The solution of equation (2.73) is easily obtained:

$$\bar{\alpha}_1(r,s) = A(r + x_0)^{-1}\exp\,[-\sqrt{s}(r + x_0)]$$

$$A = -\frac{R + x_0}{s}\exp\,[\sqrt{s}\,(R + x_0)] \tag{2.74}$$

Using (2.74) one can obtain an expression for rate constant $k_{ex}(s)$ defined by relationship (2.7):

$$k_{ex}(s) = 4\pi DR^2\exp\,(-\Phi_{eff})\left.\frac{\partial\bar{\alpha}}{\partial r}\right|_{r=R} = 4\pi D(r + x_0)\left.\frac{\partial\bar{\alpha}_1}{\partial r}\right|_{r=R}$$

The reverse Laplace transform is easily applied and the result is

$$k_{ex}(t) = 4\pi D(R + x_0)[1 + (R + x_0)/\sqrt{\pi Dt}] \tag{2.75}$$

It is clear that we obtained Smoluchowski's result (2.42) but for a new effective radius of a particle $R_e = R + x_0$. In the case of attraction potential $x_0 > 0$, value $R_e > R$ and it can be much larger than R; the non-stationarity period increases accordingly. In the case of repulsion $x_0 < 0$ (but $|x_0| < R$) the rate constant decreases, and so does the non-stationarity period.

The most interesting case is the situation of random initial distribution of particles before the reaction. To consider it, let us introduce function

$$y(r,\tau) = \alpha(r,\tau) - \alpha_{st}(r) \tag{2.76}$$

where $\alpha_{st}(r)$ is determined in (2.72).

Function $y(r,\tau)$ satisfies Equation (2.68), boundary conditions

$$y(R,\tau) = y(\infty,\tau) = 0 \tag{2.77}$$

and initial condition

$$y(r,0) = (r + x_0)^{-2}[x_0^2 - (r + x_0)(x_0 - R)] \tag{2.78}$$

Substitution $\varphi(r,\tau) = (r + x_0)y$ and the change of variable $z = r - R$ lead to a simple diffusion equation for function $\varphi(z,\tau)$:

$$\partial\varphi(z,\tau)/\partial\tau = \partial^2\varphi(z,\tau)/\partial z^2 \tag{2.79}$$

with boundary conditions

$$\varphi(0,\tau) = \varphi(\infty,\tau) = 0 \tag{2.80}$$

and initial condition

$$\varphi(z,0) = f(z) = \frac{R^2 - z(x_0 - R)}{z + (x_0 + R)} \tag{2.81}$$

The solution of problem (2.79)–(2.81) is well known, let us write the general expression

$$\varphi(z,\tau) = \frac{1}{2\sqrt{\pi\tau}} \int_0^\infty d\eta f(\eta)\{\exp[-(z-\eta)^2/(4\tau)] - $$

$$- \exp[-(z+\eta)^2/(4\tau)]\} \tag{2.82}$$

Now the rate constant can be represented in the form

$$k_{ex}(\tau) = 4\pi D(x_0 + R)\left(1 + \left.\frac{\partial\varphi}{\partial z}\right|_{z=0}\right)$$

and with use of (2.82) in the following forms

$$k_{ex}(\tau) = 4\pi D(x_0 + R)\left\{1 + \frac{1}{2\tau\sqrt{\pi\tau}}\int_0^\infty d\eta\cdot\eta\exp[-\eta^2/(4\tau)] \times\right.$$

$$\left.\times\frac{R^2 - \eta(x_0 - R)}{\eta + (x_0 + R)}\right\} = \tag{2.83a}$$

$$= 4\pi D(x_0 + R)\left\{1 + \frac{1}{2\tau\sqrt{\pi\tau}}\int_0^\infty d\eta\exp[-\eta^2/(4\tau)] \times\right.$$

$$\left.\times\left[x_0^2 - \eta(x_0 - R) - \frac{x_0^2(x_0 + R)}{\eta + (x_0 + R)}\right]\right\} \tag{2.83b}$$

Let us analyze these expressions. The reduced constant $k_{ex}(t)/4\pi DR$ is the function of dimensionless parameters Dt/R^2 and Dt/x_0^2

$$k_{ex}(t)/4\pi DR = F(Dt/R^2, Dt/x_0^2)$$

Accordingly, there are three principal regions of time change

1. $tD/R^2 \ll 1$, $Dt/x_0^2 \ll 1$ short time
2. $Dt/R^2 \gg 1$, $Dt/x_0^2 \gg 1$ large time
3. $Dt/(x_0R) \simeq 1$, cross-over region

Let us first consider the case with an attraction potential, when $x_0 > 0$. For analyzing large time at $x_0^2 \ll Dt$, it is most convenient to use Equation (2.83b) since with this limit the main contribution into the integral is from region $x_0^2 \ll \eta(x_0 - R)$. Then

$$k_{ex}(t) \simeq 4\pi R(x_0 + R)[1 - (x_0 - R)/\sqrt{\pi DT}] \tag{2.84}$$

It follows from (2.84) that the cases of strong ($x_0 \gg R$) and weak ($x_0 < R$) potentials differ considerably by the character of the rate constant $k_{ex}(t)$ approximation to the stationary value.

In the case of weak potentials

$$k_{ex}(t) = 4\pi D(x_0 + R)[1 + (R - x_0)/\sqrt{\pi Dt}]$$

which qualitatively coincides with (2.65) showing the diminution of the non-stationarity region of the rate constant with the increase of potential ($|x_0| < R$). In this case $k_{ex}(t) \geqslant k_{st}$.

The situation is quite different in the case of strong potentials. It follows from (2.84) that at $x_0 > R$ $k_{ex}(t)$ reaches the stationary value from below, i.e., $k_{ex}(t) < k_{st}$. The region of non-stationarity of the rate constant increases with the increase of the potential and covers time $t \lesssim x_0^2/D$.

Let us complement the consideration of $k_{ex}(t)$ with an analysis of short time region ($t \leqslant R^2/D$). In this limit it is convenient to use (2.83a). Assuming that the main contribution in the integral is due to the region $R^2 \gg \eta(x_0 - R)$ we obtain

$$k_{ex}(t) \simeq 4\pi D(x_0 + R)\left[1 + \frac{R^2}{(x_0 + R)\sqrt{\pi D t}}\right]$$

At $\sqrt{Dt}\, x_0 \ll R^2$, the effective rate constant $\Delta k_{ex}(t) \sim 4\pi D R^2/\sqrt{\pi D t}$ is independent on the potential, and it reproduces the Smoluchowski solution at short times.

In the time region $R^2 < (x_0 + R)\sqrt{Dt} < x_0^2$ value $k_{ex}(t)$ almost drops to value $k_D = 4\pi DR$ (which determines the stationary value of the rate constant in the absence of the interaction potential) and then at larger times rises as (2.84). Figure 2.1 illustrates the dependence of $k_{ex}(t)$.

Now let us consider the situation in the case of repulsion potentials $x_0 < 0$. The stationary value of rate constant $k_{st} = 4\pi D(R - |x_0|)$ decreases with the increase of $|x_0|$. The case of $|x_0| = R$ means an "infinite" repulsion on the boundary.

Now let $R - |x_0| = \delta < R$. It can be seen from (2.83a) and (2.83b) that $k_{ex}(t)$ monotonously changes up to the limit value $k_{st} = 4\pi D\delta$, the non-stationarity region $t \lesssim |x_0|^2/D$ growing with an increase of $|x_0|$.

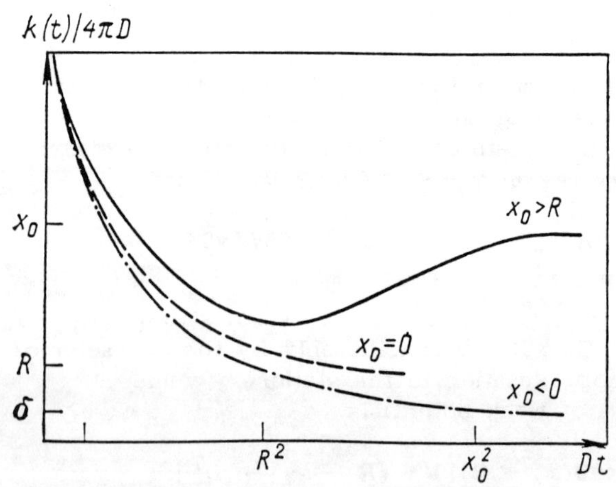

Fig. 2.1 Time dependence of "experimental" rate constant $k_{ex}(t)$.

For very strong repulsion potentials $\delta \ll R$, it is possible to give an analytic expression for $k_{ex}(t)$ at short times, but at $t \gg \delta^2 D^{-1}$:

$$k_{ex}(t) = 4\pi D\delta\left(1 + \frac{2R}{\sqrt{\pi D t}} + \frac{R^2}{2Dt}\right)$$

2.2. KINETICS OF LUMINESCENCE QUENCHING

The general expression describing the kinetics of interacting particles with generation-recombination terms taken into account can be obtained by generalizing Smoluchowski equation [112]:

$$\partial W(\vec{r},t)/\partial t = -\nabla J(\vec{r},t) + I \tag{2.85}$$

$$I = \gamma J_g W(\vec{r},t) - l_s(\vec{r})W(\vec{r},t) \tag{2.86}$$

$$\vec{J}(\vec{r}) = -D[\vec{E}(\vec{r})]\nabla W + W(\vec{r},t)\mu[\vec{E}(\vec{r})]\nabla[U(\vec{r})] \tag{2.87}$$

where J_g is the intensity of the corresponding "generating" radiation; γ is the quantum yield; the first component in the right part of (2.86) is the rate of particle generation (for example at photoionization); the second component reflects the particle losses which may be of various types, for example, quantum tunnelling [113], distant transfer of particles [114] and other processes. In the general case transfer coefficients $\mu(\vec{E})$ and $D(\vec{E})$ depend on the inner electric field $\vec{E}(\vec{r})$ determined by potential $U(\vec{r})$.

We shall consider only potentials $U(\vec{r})$ with spheric symmetry when $E(\vec{r}) = -dU/dr$, choosing as a rule the power-type dependence for $U(\vec{r})$:

$$U(r) = Ar^{-m} \tag{2.88}$$

For purely ionic reactions, when $m = 1$, value A is defined by the dielectric permittivity of the medium and the charges of the corresponding ions.

In reactions between neutral particles, the interaction has an induction character and is described by expression (2.88) at $m = 6$, A being dependent on the polarizability and dipole moments of molecules.

It should be noted that the usual real potential of interaction between components of the system is unknown. In the general case, it can be found (together with the solution of the general transfer problem, see Section 2.1) using some self-consistent procedures, applying for example the Debye-Hückel method which leads to exponential potentials

$$U(r) = \frac{a}{r+b}\exp(-rd)$$

where d^{-1} is just the Debye radius (screening radius); a and b are parameters defining the unscreened potential.

2.2.1. Uncharged Particles

Let us first consider the simplest case of interacting particle kinetics when D = const, $\mu(E)$ = const, $U(r) = 0$. Obviously the usual "two-parameter" model of traditional DCR kinetics [114] dealing with only values k and R forming k_{ex} can describe a limited group of reactions where diffusion transfer may play a significant role. Indeed, this model does not consider the possible particle interactions at distances larger than R. Anisotropy of particles is ignored and the interaction of reagents and the dynamics of their motion at the moment of contact are not taken into account. Let us discuss these questions in more detail.

The motion of particles up to their contact ($r \gg R$) is usually described as diffusional. However, the short-range order in liquids may considerably change the character of the motion at various distances (comparable to the size of the particles of the solvent). The formation of solvate shells of reagents increases their difference.

Traditional DCR kinetics does not take into consideration the possibility of finding colliding particles in various states (conformational, orientational), which can also noticeably change the calculated value of the rate constant. The possible reaction of diffusing components long before their direct contact (e.g., in the case of resonance energy transfer because of long-range interaction or tunnelling) should also be taken into account.

Let us discuss the latter of the above-mentioned cases and try to exceed the limits of the concept used in DCR kinetics of reaction radius. For definition, it is convenient to consider the problem of luminescence quenching and the theory of diffusion accelerated distance transfer. The process is formalized in the following way. Let there be two types of molecules A and B in the system which are capable of excitation, and let the excitation be passed from A^* to another type of molecule B, but reverse energy transfer is prohibited. Schematically this process can be presented as

$$B + A^* \longrightarrow B^* + A$$

i.e., each type B molecule performs the role of a trap of infinite capacity for active particles A^*. If the process took place only upon the collision of B with A^*, the situation would be typical of the Smoluchowski method which was first applied in working out the theory of luminescence quenching. However such analysis gives very large (as compared to molecular) values of R_e capture radius. Besides, what is more important, with an increase of the diffusion coefficient the calculated value of R_e decreased while that of k_{ex} increased. The usual expressions $k_{ex} = kk_D/(k + k_D)$ and $R_e = (k_{ex}/k)R$ do not explain such behavior and this has led to the theory of distant luminescence quenching [4].

If the excitation source is "turned off" the main equation

$$\partial n(r,t)/\partial t = D\Delta n(r,t) - \overline{U}(r)n(r,t) \tag{2.89}$$

can be used instead of (2.85)–(2.87). Here $n(r,t)$ is the concentration of excited molecules A; $\overline{U}$ is the local probability of quenching.

When imposing boundary conditions it is assumed that upon collision no new channels of quenching appear and this yields condition

$$dn/dr\Big| = 0 \tag{2.90}$$

At large distances from the quencher, excitation density is considered constant

$$n(\infty, t) = 1 \tag{2.91}$$

and equal to the initial concentration of molecules

$$n(r, 0) = 1 \tag{2.92}$$

Integrating Equation (2.89) over the volume, one can obtain the excitation flux into the quenching center which is equivalent to the introduction of rate constant k_q of the luminescence quenching process:

$$k_q = \int \overline{U}(\vec{r})n(\vec{r}, t)d\vec{r} = 4\pi \int_R^\infty \overline{U}(r)n(r, t)r^2 dr \tag{2.93}$$

The kinetics of luminescence quenching is determined by the equation for mean (over volume) excitation concentrations $N(t)$

$$N(t) = \frac{1}{V} \int n(r, t)d\vec{r} \tag{2.94}$$

$$dN/dt = -k_q(t)cN(t) \tag{2.95}$$

where c is the concentration of the quenching impurities.

Value k_q, being in the general case a function of time, depends on parameters defining interaction $\overline{U}_r(\vec{r})$. For the usually considered interactions—dipole-dipole and the exchange one—we obtain respectively

$$\overline{U}_d = \alpha_u/r^6 \tag{2.96}$$

$$\overline{U}_{ex} = \overline{U}_0 \exp(-2xr) = U_m \exp[-2x(r - R)] \tag{2.97}$$

where parameter α_u determines the quenching probability per time unit for dipole-dipole interaction; $\overline{U}_m$ is the quenching probability per time unit in the case of exchange interaction at maximal approaching of reagents; x^{-1} is the size characterizing the extent of reactant wave function overlap.

While discussing the type of k_q, it is necessary to solve non-stationary Equation (2.89) which is quite difficult. However, it is clear that in the case of slow diffusion, when the excitation density in the thin layer in the vicinity of the quencher has time to exhaust itself due to the quenching process [the second component in the first part of (2.89)], the whole process will be limited by the diffusion approximation of reagents. Then k_q is analogous to the usual constant k_D but with a new capture radius

$$k_q = 4\pi D R_e \tag{2.98}$$

Using (2.96) and the stationary solution of (2.89) with condition (2.90–2.92), one can write the expression for R_e in explicit form [115]:

$$R_e = \left(\frac{\alpha}{D}\right)^{1/4} \frac{2\Gamma(3/4)}{\Gamma(1/4)} I^{-1}_{-3/4}\left(\frac{1}{2R^2}\sqrt{\frac{\alpha_u}{D}}\right) I_{3/4}\left(\frac{1}{2R^2}\sqrt{\frac{\alpha_u}{D}}\right)$$

where $\Gamma(x)$ is the gamma function; $I_{\pm 3/4}$ are modified Bessel functions [116].

Let us consider the limit cases.

In the first case let $R^{-4}D^{-1}\alpha_u \ll 1$. Since

$$I_\nu(x) \simeq 2^{-\nu} x^\nu \Gamma^{-1}(1 + \nu), \qquad \Gamma(1 + z) = z\Gamma(z)$$

then within this limit

$$R_e \simeq \frac{\alpha_u}{D} \frac{1}{3R^3} \ll R \tag{2.99}$$

The inequality obtained points to the limitation of the model as $R_e \ll R$ and the model of the thin layer [114] is invalid. However, in this case the rate constant ceases being dependent on the diffusion coefficient:

$$k_q = 4\pi D \frac{\alpha_u}{3DR^3} = \frac{4\pi\alpha_u}{3R^3}$$

This is a result equivalent to the assumption about the constancy of excitation concentration in the vicinity of the quencher: $n(r, t) = n(r, 0) = 1$ [117].

Now let us analyze the opposite case of slow diffusion, when $\alpha_u/(R^4 D) \gg 1$. Using the asymptotic representation of modified Bessel functions $I_\nu(x)$ at large values of the argument $x \to \infty$, namely

$$I_\nu(x) \simeq \exp(x)/\sqrt{2\pi x}$$

we obtain

$$R_e = 2\frac{\Gamma(3/4)}{\Gamma(1/4)}\left(\frac{\alpha_u}{D}\right)^{1/4} \tag{2.100}$$

i.e., R_e is not at all dependent on the size of the colliding particles and the process is determined only by parameter α_u, characterizing the intensity of the transfer process and diffusion coefficient D. At this

Thus, in the case of slow diffusion ($\alpha_u/(R^4D) \gg 1$, i.e., $R_e{}^4/R^4 \gg 1$) it is possible to describe luminescence quenching applying the Smoluchowski method with a new capture radius ($R \to R_e$). Then rate constant k_q should reach a steady value at $t \gg t_1$, where $t_1 \simeq R_e{}^2/D$ is the time of diffusion transport through the region in which the probability of distant excitation transfer is large. At usual (for liquid) parameters [4] $\alpha_u \simeq (10^{-33} - 10^{-31})$ cm^6/s and $D \simeq 10^{-5}$ cm^2/s one can obtain

$$\frac{\alpha_u}{D} = (10^{-28} - 10^{-26})\ \text{cm}^4, \qquad \left(\frac{\alpha_u}{D}\right)^{1/4} \sim 1\ \text{nm}$$

at $R \simeq (0.7 - 0.9)$ nm [20] we obtain $(R_e - R)/R \ll 1$ ($t_1 \simeq 10^{-9}$ s). Thus it is possible to speak about the model of a thin quenching layer [114] in which the extensive interaction of reagents can be replaced by a step of equivalent thickness $(R_e - R)$ and in the analysis of the kinetics of the process the Smoluchowski method can be used. In this event, the application of the distance model to quenching problems provides for establishing the microscopic sense of the parameters of the traditional model which is treated as a phenomenological one. A detailed discussion of the distance model can be found in reviews [114,118,119].

Note that the given statement of the problem does not take into account correlation effects—a deficiency inherent in the Smoluchowski method. The formal structure of (2.89) is analogous to the general form of the equation with correlations taken into account, the same as the physical sense of the notion of distance interaction essentially reflects (in a certain form) correlation effects. A more detailed analysis of this question which reviews a great amount of experimental data can be found in [119]. The simplest way to understand the role of correlations is to consider the same theory (distance transfer) but with an internal field—potential $U(r)$ between reacting charged particles, for example.

2.2.2. Charged Particles

Experimentally and theoretically the processes of energy transfer upon particle excitation carrying a charge are analyzed in [120–123]. A discussion of this problem brings us back to the Smoluchowski equation including in the right part a term of the sink connected with the distance transfer, i.e., back to Equations (2.85–2.87) at $J_g = 0$, $l_s = \overline{U}(r)$. As has already been mentioned (see Section 2.1), the introduction of interaction potential $U(r)$ changes the rate of the chemical transformation

$$k(U) = k(0) \exp[-\beta U(R)] \tag{2.101}$$

and the radius of the reaction sphere by substituting R for Debye radius R_D (in considering the stationary Smoluchowski equation). Taking into account the distance transfer of excitation introduces additional aspects into the understanding of the kinetics of the processes (see [124]). Let us discuss these new aspects choosing as an example Coulomb particle interaction [see (2.9)]:

$$U(r) = -q_0/r \tag{2.102}$$

We shall consider stationary equation (2.85)

$$D\frac{1}{r^2}\frac{d}{dr}\left[r^2\left(\frac{dn}{dr}+\frac{n}{\beta}\frac{dU}{dr}\right)\right]-\bar{U}n = 0 \tag{2.103}$$

with boundary conditions

$$n(\infty) = 1, \quad \left(\frac{dn}{dr}+\frac{n}{\beta}\frac{dU}{dr}\right)\bigg|_{r=R} = 0 \tag{2.104}$$

Rate constant (2.93) can now be rewritten in a form analogous to (2.98), introducing capture radius R_Q for the case when the reagents are linked by the interaction potential $k_q = 4\pi DR_Q$. Here

$$R_Q = D^{-1}\int_R^\infty \bar{U}nr^2 dr = \lim_{r\to\infty} r^2\left(\frac{dn}{dr}+n\beta\frac{dU}{dr}\right)$$

To solve (2.103) it is convenient to make the substitution

$$g(r) = n(r)\exp(\beta U)$$

which in the case of Coulomb potential (2.102) gives the equation

$$g'' + g'\left(\frac{2}{r}+\frac{\bar{\beta}}{r^2}\right) - \alpha_u r^{-6} g = 0 \tag{2.105}$$

where $\bar{\beta} = -\beta q_0$.

Function $g(r)$ must satisfy boundary conditions

$$\frac{dg}{dr}\bigg|_{r=R} = g'(R) = 0, \qquad g(\infty) = 1 \tag{2.106}$$

In this case the expression for determining the capture radius assumes the simple form

$$R_Q = \lim_{r\to\infty}(r^2 dg/dr)$$

It is convenient to pass on to dimensionless variable $x = R/r$. Then (2.105) can be presented in the form

$$\frac{d^2g}{dx^2} - 4b\frac{dg}{dx} - 16\beta_m x^2 g(x) = 0 \tag{2.107}$$

with boundary conditions

$$g(0) = 1, \qquad g'|_{x=1} = 0 \tag{2.108}$$

Parameters b and β_m in (2.106) play the principal role in determining the regimes of the quenching process and have the form

$$b = q_0\beta/(4R), \qquad \beta_m = \alpha_u/(16DR^4)$$

It has been shown in Section 2.21, that at $\beta_m \gg 1$, it becomes possible to describe the process using traditional theory with a boundary condition of the third kind, i.e., in the model of a "grey sphere". Here it is interesting to determine how the turning-off of electrostatic interaction will manifest itself.

The possible cases can be discussed introducing parameter R_s with the help of function $\beta_m = \alpha_u/(16\pi Dr^4)$, requiring that $\beta_m(R_s) = 1$. Then parameter

$$R_s = \frac{1}{2}\left(\frac{\alpha_u}{D}\right)^{1/4}$$

characterizes the distance beyond which ($r > R_s$) the quenching is weak, whereas within this distance ($r < R_s$)—strong. The characteristic influence region of potential $U(r)$ is defined by the Onsager radius $r_{On} = |q_0|\beta$. Apparently the strongest effects should be expected in the case of a repulsive potential ($b > 0$) when Coulomb interaction hinders close particle convergence. Then $R > R_s > r_{On}$ should be chosen.

As has already been shown, at $\beta_m \gg 1$ value k_q becomes independent of R and effective radius R_e is determined only by the character of interaction $\overline{U}$. In this situation, taking into account field $U(r)$ can lead to different consequences. A weak field when $r_{On} \simeq R$ will not lead to new effects. However, if field $U(r)$ is repulsive and $r_{On} \gg R$, the particles can approach each other up to distance r_{On}. If $R_s \ll r_{On}$, the situation becomes similar to the one realized in the case of a weak quenching regime, since $\beta(r_{On}) \ll 1$. The special case is quite difficult for analysis and determined by the relationship

$$R_s \simeq r_{On}$$

Let us first consider region $R \ll R_s \ll r_{On}$ ($\beta_m \gg 1$) and use the approximate procedure suggested in [114]. We shall seek the solution of Equation (2.105) using the method of successive approximations:

$$g(x) = g^{(0)}(x) + g^{(1)}(x) + g^{(2)}(x) + \ldots$$

assuming the solution of the equation

$$g_0'' - 4bg_0' = 0, \qquad g_0 = g^{(0)}$$

to be the zero approximation at $g_0(0) = 1$, $g_0'(1) = 0$ [condition (2.108)]. It is evident that $g_0(x) = \text{const} = 1$. The rest of the functions can be determined from the iterative equation

$$g^{(i)''} - 4bg^{(i)'} - 16\beta_m x^2 g^{(i-1)} = 0$$

with boundary conditions (2.108) for each $g^{(i)}$. The thus sought after function $g(x)$ satisfies boundary conditions (2.108) and is the solution of Equation (2.107). In particular, for $g_1 = g^{(1)}$ we obtain

$$g_1'' - 4bg_1' - 16\beta_m x^2 = 0$$

which leads to an expression for R_Q of the following type

$$R_Q = (2b^3)^{-1}\beta_m R[1 - (8b^2 + 4b + 1)\exp(-4b)]$$

At $b \gg 1$ we get

$$R_Q = \frac{\beta_m R}{2b^3} = \frac{2\alpha_u}{Dr_{On}^3} < r_{On} \qquad (2.109)$$

Therefore result (2.109) is analogous to expression (2.99) (upon substitution $R \to r_{On}$). At the same time, since R_Q is independent of R, the property of "strong" interaction (or diffusion control of the reaction) is preserved [analogous to (2.100)] resulting from inequality $\beta_m \gg 1$. Thus repulsive interaction in this case changes the limiting stages (from diffusion al—to chemical), increasing the effective radius of the particle volume proper.

A more detailed account of this problem, including consideration of the quenching kinetics at different relations of values R, R_s and r_{On}, can be found in the works of Burshtein, Doktorov, etc. [124].

We shall only make two observations. First, the application of the quasi-stationary approximation for calculating k_q = const with the following analysis of non-stationarity using (2.95), generally speaking, permits analyzing only the deep stages of the reaction [therefore case $k_q = k_q(t)$ was not analyzed]. Secondly, using assumption (2.92) with this type of initial distribution should be problematic since the initial distribution of charged [or, in general, interacting by means of potential $U(r)$] particles is correlated.

Both these questions are widely discussed in an analysis of a related phenomenon—recombination luminescence [101] which accompanies radiolysis or photolysis of solid organic systems.

It has been established that recombination radiation is a direct result of energy release in particle recombination of the opposite sign (both free and bound), the radiation intensity being proportional to the disappearance rate of particles in the recombination process ([125,126] on anomalies [127]) Let us enumerate the research directions in which this method is being successfully employed:

1. Investigation of the recombination mechanism of charges and determination of the contribution of diffusion motion and tunnelling to this process;

2. Influence of the character of the medium on the ionization potential of impurities;

3. Study of the mechanism of elementary reactions resulting in photon radiation;

4. Investigation of the effect of different factors (temperature, electric field, light) upon the recombination process [101].

Among the experimentally observed dependences of recombination radiation intensity on time after UV-radiation are:

$$L(t)/L(0) = [1 + k_1(t - t_0)]^{-1}$$

and (the more interesting for us)

$$L \sim kt^{-m}$$

where k_1, k, m are parameters, and $m \simeq 1.0$ [128].

If recombination is limited by the preceding diffusion motion of charged particles, the corresponding temperature dependences (resulting from the activation character of the diffusion transfer) should be experimentally observed. However, there is a great amount of experimental data [101, 129], which contradict these consequences of diffusion theory. Now prevalent is the viewpoint that the charge recombination especially at low temperatures, is the result of the tunnelling of an electron from the anion on to the cation. It is assumed that the tunnel transfer is achieved at considerable distances (up to 5 nm) [101].

2.3. TUNNEL CHARGE TRANSFER IN RECOMBINATION REACTIONS [130–136]

Let us consider the generalized Smoluchowski equation (2.85–2.87) at $J_g = 0$, $U(r) = 0$ to be the initial one:

$$\frac{\partial n(r,t)}{\partial t} = -l_s{}^q n(r,t) + D\Delta n(r,t) \tag{2.110}$$

with sink function $l_s{}^q$ depending on the probability of tunnel particle transfer per time unit. It is quite often assumed that

$$W_{\text{tun}} = \nu_0 \exp(-2r/R) \tag{2.111}$$

i.e., has the same form as in the theory of distant luminescence quenching at exchange interaction between the donor and acceptor.

Generally the approach widely used is to separately consider the distribution of donors and acceptors, which are described by conditional concentrations of donors $n(r,t)$ and acceptors $N(r,t)$ averaged at time t over the ensemble of particles of opposite type

$$n(r,t) = \sum_{i=1}^{N(t)} n_i(r,t)$$

$$N(r,t) = \sum_{i=1}^{n(t)} N_i(r,t) \qquad (2.112)$$

where $n_i(r,t)[N_i(r,t)]$ is the concentration of donor (acceptor) particles located at distance r from the arbitrarily chosen i-th acceptor (donor) particle; $n(t)$ and $N(t)$ are the mean over-the-volume concentration of donors and acceptors.

Conditional and mean-by-volume concentrations are connected by relationships [113]:

$$n(t) = \lim_{r\to\infty} n(r,t)$$

$$N(t) = \lim_{r\to\infty} N(r,t)$$

$$n(t)N(r,t) = N(t)n(r,t); \quad n(0) - n(t) = N(0) - N(t)$$

where $n(0), N(0)$ are the mean initial concentrations.

Then the changes of $n(r,t)$ and $N(r,t)$ are described by a system of equations of type (2.110) where

$$l_s^q(n) \quad \left[W_{tun}(r) + \int_0^\infty W_{tun}(\rho)4\pi\rho^2 N(\rho,t)d\rho\right] \qquad (2.113)$$

i.e.

$$\frac{\partial n(r,t)}{\partial t} = -n(r,t)l_s^q(N) + \frac{D}{r^2}\frac{\partial}{\partial r}\left(r^2\frac{\partial n}{\partial r}\right)$$

$$\frac{\partial N(r,t)}{\partial t} = -N(r,t)l_s^q(n) + \frac{D}{r^2}\frac{\partial}{\partial r}\left(r^2\frac{\partial N}{\partial r}\right)$$

For the mean concentration one obtains the equations

$$\frac{dn}{dt} = -n(t)\int_0^\infty W_{tun}(\rho)4\pi\rho^2 N(\rho,t)d\rho$$

$$\frac{dN}{dt} = -N(t) \int_0^\infty W_{tun}(\rho) 4\pi\rho^2 n(\rho, t) d\rho \tag{2.115}$$

It is easily seen that system (2.114), (2.115) is practically speaking the generalization of the formalism of the distance quenching theory, and it is reduced to the above-considered equations if N = const and the transfer of an electron anywhere except to the acceptor molecule is prohibited, i.e., the first term in (2.113) is ignored. The general solution of this system is found in [136]. For perhaps the most important case of random distribution of reagents and $N(0) \gg n(0)$, it is possible to write the solution for two limit situations: at $t \gg \tau_D$ and $t \ll \tau_D$ where $\tau_D = R^2/D$ is the time of the diffusion displacement of reagents at the distances of about R, treated as the characteristic parameter of the DCR theory using the Smoluchowski method. As it should be expected at $t \gg \tau_D$, the exponential dependence for $n(t)$ is obtained

$$n(t)/n_0 = \exp\left[-\frac{4}{3}\pi R_\tau^3 N(0) - 4\pi D N(0) t\right] \tag{2.116}$$

where

$$R_\tau = \frac{1}{2} R \ln(\nu\tau_0), \qquad \nu = W \exp(2r/R)$$

This yields the effective size of particles ($R \rightarrow R_\tau$) now described by value R_τ. This value is the particular case of the general expression determining distance R_t to which the electron tunnels in time t, i.e., R_τ is the tunnelling distance in time τ_D. Value R_τ plays the principal role in the description of the reaction kinetics at short times.

At $t \ll \tau_D$, the rate of the tunnel reaction under the condition that $N(0) \gg n(0)$ is described by the expression

$$n(t)/n(0) = \exp\left[-\frac{\pi R}{6} N(0) \ln^3 \nu t\right] \tag{2.117}$$

If $N(0) = n(0)$ the corresponding expression has the form

$$n(t)/n(0) = \left[1 + \frac{\pi R^3}{6} n(0) \ln^3 \nu t\right]^{-1} \tag{2.118}$$

Value

$$\exp\left[-\frac{\pi R^3}{6} N(0) \ln^3 \nu t\right] \equiv \exp\left[-\frac{4\pi}{3} N(0) R_t^3\right]$$

is the probability for a donor particle to appear outside the effective reaction volume $V_t = 4/3\, \pi R_t^3$ near an arbitrary acceptor particle, i.e., the probability that the particle will not be annihilated in the course of a tunnel reaction by time t.

The application of relationships (2.116–2.118) in a number of cases provided for a quantitative description of the experiment [113].

2.4. MODEL POTENTIALS IN THE INTERACTING PARTICLE KINETICS [95]

We have considered several processes in which a significant role is played by the interaction between active centers. However, in the general case the Smoluchowski equation (2.85–2.87) provides a basis for an investigation of only stationary (or quasi-stationary) stages of these processes since the non-stationary solutions of this equation are practically unknown.

The difficulty of analyzing non-stationary solutions of the Smoluchowski equation can be avoided either with the aid of a numerical solution or by employing methods based on different techniques of matching approximate solutions [103]. However, the problem then appears of controlling the results in both the above-mentioned approaches.

In this connection, a reverse problem can be posed, i.e., to seek model potentials permitting an anlytical solution of the Smoluchowski equation. This statement of the problem is also justified by the fact that the real potentials of interaction between particles are unknown even in the absence of chemical reactions. They are generally effective and must be determined through some self-consistent procedure, as for example, in the Debye-Hückel method. The course of chemical reactions can considerably complicate the search for the effective potentials, therefore the type of potential is usually chosen using physical considerations, and the use of model potentials is quite permissible.

The thus-obtained analytic solutions can be applied in analyzing the kinetic regularities of chemical processes. In particular, for finding the rate constants of chemical reactions which generally depend on time and type of the potential. Let us consider the possible types of such potentials. For simplicity, we shall consider only three-dimensional equation (2.2) and write the equation for function $Z(r,t) = rW(r,t)\exp[\frac{1}{2}\beta U(r)]$:

$$\frac{1}{D}\frac{\partial^2 Z(r,t)}{\partial t} = \frac{\partial^2 Z}{\partial r^2} + F(r)Z \tag{2.119}$$

where

$$F(r) = \frac{1}{2}V'' + r^{-1}V' - \frac{1}{4}(V')^2, \qquad V = \beta U(r) \tag{2.120}$$

To seek model potentials, it is convenient to assume that

$$V(r) = -2\ln\eta, \qquad \frac{1}{2}V' = -\eta'/\eta \tag{2.121}$$

which upon substitution into (2.120) yields

$$\eta'' + \frac{2}{r}\eta' + \eta F(r) = 0 \qquad (2.122)$$

Condition (2.122) can be treated as an equation, i.e., function $F(r)$ giving realistic potentials can be either chosen or considered known (instead of the usual method of discussing the problem when either η or V are known). The boundary condition for (2.122) follows from (2.121):

$$\eta(r) = 1 \quad \text{at} \quad r \longrightarrow \infty \qquad (2.123)$$

Let us now consider the possible cases of solutions of Equation (2.122) which at the same time provide for solving Equation (2.119). Applying the Laplace transformation transforms Equation (2.119) to the type

$$\frac{d^2\bar{Z}(r,s)}{dr^2} + \bar{Z}(r,s)[F(r) - s] = -Z(r,0)$$

where

$$\bar{Z}(r,s) = \int_0^\infty e^{-s\tau} Z(r,\tau)d\tau, \qquad \tau = Dt$$

It can be seen that general equation (2.119) can be solved analytically if equation

$$d^2\nu(r)/dr^2 + \nu(r)F_1(r) = 0 \qquad (2.124)$$

obtained from (2.122) upon introducing function $\nu(r) = r\eta(r)$, $F_1 = F +$ + const. is analytically soluble. Function $F_1(r)$ then differs from $F(r)$ in (2.120) by a value independent of r. For potential $V_{eff} = -2\ln\eta$ to tend to zero at $r \to \infty$ function $\nu(r)$ at $r \to \infty$ must have the form $\nu = a + r$. This means that function $F(r)$ should tend to zero faster than $(r^2 + \varepsilon)^{-1}$ (where $\varepsilon > 0$). A number of functions $F(r)$ satisfy this condition:

1. $F(r) = 0$ (2.125)

2. Function of the Morse type

$$F(r) = A_1 \exp(-\lambda r) + A_2 \exp(-2\lambda r) \qquad (2.126)$$

3. Function of the Ekkart type

$$F(r) = A_1[1 + \exp(\lambda r)]^{-1} + A_2[1 + \exp(2\lambda r)]^{-2} \qquad (2.127)$$

Now let us discuss the simplest cases. Condition (2.125) gives

$$\eta(r) = 1 + x_0/r, \qquad V_{\text{eff}} = -2\ln(1 + x_0/r) \tag{2.128}$$

(where x_0 is a parameter). The behavior of potential (2.128) at large distances $r \gg x_0$ is Coulomb like:

$$\beta U_{\text{eff}} = V_{\text{eff}} \simeq -2x_0/r$$

Parameter x_0 can be expressed through the parameters of Coulomb potential U_{Coul}:

$$\beta U_{\text{coul}}(r) = \beta\,\frac{q_1q_2}{\varepsilon r} = \beta_0\,\frac{1}{r}, \qquad \beta_0 = \frac{q_1q_2}{\varepsilon}\,\beta$$

In the event of attraction one obtains ($x_0 > 0$):

$$\beta U_{\text{coul}}(r) = -\beta_0/r \simeq V_{\text{eff}} = -2x_0/r, \qquad x_0 = \beta_0/2$$

The case $x_0 < 0$ corresponds to the repulsion and $|x_0| < R$.

Now let function $F(r)$ be determined by an expression of type (2.126). Let us assume that

$$F(r) = -c^2[\nu^2 + b^2\exp(2cr)], \qquad c < 0 \tag{2.129}$$

Now we come to variable g

$$g = b\exp(cr), \qquad g' = cg, \qquad g'' = c^2 g \tag{2.130}$$

and use (2.130) to transform (2.124) and (2.129):

$$\frac{d^2\nu(g)}{dg^2} + \frac{1}{g}\,\frac{d\nu(g)}{dg} - \nu(g)\left[1 + \frac{\nu^2}{g^2}\right] = 0$$

We obtain the Bessel equation for functions of the imaginary argument. Its solution has the form

$$\nu(g) = A_1 I_\nu(g) + A_2 K_\nu(g) \tag{2.131}$$

Now using (2.131), (2.129) and (2.121) we can derive

$$V_{\text{eff}} = -2\ln\left[r^{-1}A_1 I_\nu(be^{-\sigma r}) + r^{-1}A_2 K_\nu(be^{-\sigma r})\right], \qquad \sigma = |c| \tag{2.132}$$

Let us consider the asymptotic behavior of V_{eff} at $r \to \infty$ when $g \to 0$.

Since

$$I_\nu(g) \simeq 2^{-\nu} g^\nu \Gamma^{-1}(1+\nu), \qquad K_\nu(g) \simeq \begin{cases} 2^{\nu-1} g^{-\nu} \Gamma(\nu), & \nu > 0 \\ \ln \dfrac{2}{g}, & \nu = 0,\ g \to \infty \end{cases}$$

the first term of square brackets (2.132) always tend to zero, the second—to a constant value only at $\nu = 0$, when combination $r^{-1} A_2 K_0(g) \to A_2 \sigma$. Assuming that $A_2 = 1/\sigma$ one can determine the "family" of potentials:

$$V_{\text{eff}}(r) = -2 \ln \left[\frac{A_1}{r} I_0(be^{-\sigma r}) + \frac{1}{\sigma r} K_0(be^{-\sigma r}) \right] \tag{2.133}$$

For analysis, expression (2.133) can be presented somewhat differently using the relationship [137]:

$$K_0(g) = -\left(\ln \frac{g}{2} + \gamma\right) I_0(g) + \frac{g^2}{4(1!)^2} + \frac{(g^2/4)^2}{(2!)^2}\left(1 + \frac{1}{2}\right) + \ldots$$

where γ is the Euler constant. Then for large value of r ($g \ll 1$) we get

$$V_{\text{eff}} \simeq -2 \ln \left\{ I_0(be^{-\sigma r}) + r^{-1} I_0(be^{-\sigma r}) \left[A_1 - \left(\gamma + \ln \frac{b}{2}\right) \sigma^{-1} \right] + \right.$$
$$\left. + 0(e^{-2\sigma r}/r) \right\} \tag{2.134}$$

Two groups of potentials with different asymptotic behavior at infinity ($g \to 0$) follow from (2.133) and (2.134).

Let $A_1 - [\gamma + \ln (b/2)]\sigma^{-1} = 0$. Then

$$V_{\text{eff}} \simeq -2 \ln \left[1 + \frac{b^2}{4} \exp(-2\sigma r) \right] \simeq -\frac{1}{2} b^2 \exp(-2\sigma r)$$

i.e., we derived extent potential exponentially decreasing at $r \to \infty$. These potentials contain two adjustable parameters and provide for an investigation of non-stationary solutions of the Smoluchowski equation at weak and strong potentials in the vicinity of the particle reaction surface.

It can be easily seen that at $A_1 - [\gamma + \ln (b/2)]\sigma^{-1} \neq 0$ potentials (2.133) decrease in inverse proportion to the distance, i.e., they have the asymptotics of Coulomb type.

CHAPTER 3

METHODS OF MANY-PARTICLE DESCRIPTION IN THE KINETICS OF DIFFUSION CONTROLLED PROCESSES

3.1. MANY-PARTICLE DISTRIBUTION FUNCTIONS

The previous chapters discuss the principles of traditional diffusion controlled kinetics. Its undeniable merits are simplicity, qualitative "correctness", and the fact it can be easily employed and explained on the basis of a great amount of experimental data [8,15,96,138,139].

At the same time, the largely informative character of present day experiments gave rise to new problems which are quite difficult to analyze within the limits of traditional kinetics. Let us take, for example, the problem of the "capture" of elementary excitations at randomly distributed static traps. For this problem the Smoluchowski method predicts exponential decreasing of excitation concentrations at large times. However, quite often the observed dependences do not conform to the simple exponential law [140]. Exceeding the limits of the Smoluchowski method is required in analyzing questions connected with an experimental study by quasi-elastic light scattering methods [176] of molecular motions of a medium and their effect upon fluctuation formation of reagent concentrations. All this has stimulated further work towards reconsidering the "conceptual" principles of the traditional theory [16,75] and exposed its limitations and deficiencies.

1. The main limitation is connected with the use of the diffusion approximation which is valid only at low concentrations of components. The fluctuations of reagent concentrations close to the reaction surface are usually ignored in this case because of the competition between different-type particles for a place in the reaction cage [16] in the course of chemical transformation.

2. The dynamics of the elementary act of chemical transformation are almost not involved in the analysis; the theory does not discuss the problem of the interrelation of direct and reverse processes which should be reflected [31,175] in the corresponding phenomenological relations between the rate constants in boundary conditions.

3. The nature of the molecular mobility determined by the many-parameter interaction between the particle and the medium is not specified.

Only an analysis of elementary act mobility based on successively introducing the system Hamiltonian permits, in principle, an evaluation of the adequacy of introducing one phenomenological diffusion coefficient for describing macrotransport in the system (see Section 1.9 and [75]).

4. It is assumed that asymptotic dependences of concentrations at large times are defined by the equations of formal kinetics. However, the kinetics of establishing complete equilibrium in the system can, to a considerable extent, be determined by the fluctuations of concentrations [12].

The development of present-day theory provides for discussing these problems. Let us consider some of the results. Taking into account correlations in particle positions results in a generalization of the Smoluchowski method. Construction of the general theory of diffusion controlled reactions entails the introduction of multiparticle distribution functions and adequate boundary conditions [16,141–144]. Let the system consist of N particles and let $C^N(\vec{r}_1,\vec{r}_2,\ldots,\vec{r}_N,t)$ be the N-particle distribution function having the meaning of probability density for finding the first particle in the volume $d\vec{r}_1$ at point $\vec{r}_1$, the second particle—in the volume $d\vec{r}_2$ at point $\vec{r}_2$, etc., at time t. It is convenient to normalize C^N distribution functions to 1:

$$\int C^N(\vec{r}_1,\vec{r}_2,\ldots,\vec{r}_N,t)\,d\vec{r}_1 d\vec{r}_2\ldots d\vec{r}_N = 1 \tag{3.1}$$

Distribution functions of lower order n are obtained from partial integration of function C^N:

$$C^n(\vec{r}_1,\vec{r}_2,\ldots,\vec{r}_n,t) = \int C^N(\vec{r}_1,\vec{r}_2,\ldots,\vec{r}_N,t)\,d\vec{r}_{n+1}\ldots d\vec{r}_N \tag{3.2}$$

If no chemical reactions occur in the system and the particles move independently, the introduced n-particle distribution function obeys diffusion equation:

$$\frac{\partial C^n(\ldots\vec{r}_j,\ldots,t)}{\partial t} = \sum_{j=1}^{n} \nabla_j D_j \nabla_j C^n(\ldots,\vec{r}_j,\ldots,t) \tag{3.3}$$

where ∇ is the gradient operator; and D_j is the diffusion coefficient of an j-particle.

We shall assume that D_j = const. Equation (3.3) allows the separation of variables which is equivalent to the relation

$$C^n(\vec{r}_1,\ldots,\vec{r}_n,t) = C_1(\vec{r}_1,t)C_2(\vec{r}_2,t)\ldots C_n(\vec{r}_n,t) \tag{3.4}$$

In the equilibrium state

$$C_j = V_0^{-1} = \text{const}, \qquad C^n = V_0^{-n} \tag{3.5}$$

where V_0 is the total volume of the system.

Expressions (3.4) and (3.5) are characteristic of systems with random particle arrangement. We earlier used such a distribution in discussing the Smoluchowski method. Generally the chemical transformation of particles leads to concentrational inhomogeneity of the system and particle distribution appears to be "non-random", which requires consideration of more general than one-particle distribution functions. Let us point out some properties of random distribution.

a) Probability $P_v^{(0)}$ of volume $V < V_0$ not containing particles (the volume is free of particles is equal to

$$P_v^{(0)} = \int\limits_{v_0 - v} C^N(\vec{r}_1,\ldots,\vec{r}_N, t)\,d\vec{r}_1\ldots d\vec{r}_N = \left(1 - \frac{V}{V_0}\right)^N \simeq e^{-V\rho_0} \tag{3.6}$$

where $\rho_0 = N/V_0$ is the particle density in the system.

Formula (3.6) is called the Perrin formula.

b) Probability $P_v^{(k)}$ of volume V containing one, two, etc., particles is equal to

$$P_v^{(k)} = \frac{(V\rho_0)^k}{k!}\, e^{-V\rho_0}, \qquad k = 1;\ 2;\ldots \tag{3.7}$$

This expression is called the Poisson distribution.

Smoluchowski's analysis is based on considering a one-particle distribution function. The analysis can be made more complete by introducing two-particle functions according to rule (2.2):

$$C_{ij}^{(2)}(\vec{r}_i, \vec{r}_j, t) = \int C^N(\vec{r}_1,\ldots,\vec{r}_N, t) \prod_{k \neq i,j}^{N} d\vec{r}_k \tag{3.8}$$

The analysis of equations derived using functions (3.8) can be found in papers [22,23,145,146].

However, it is more convenient to deal with two-particle functions defined somewhat differently [147,148]. Let $\vec{r}$ be the radius-vector between any two particles. We shall select one of the particles in the system. Function $\rho(\vec{r}, t)$ shall be determined so that ρdV is the probability of detecting a particle located at distance r from the selected one in volume dV at time t. These functions will be introduced more rigorously below, whereas here we shall confine ourselves to a qualitative consideration. In the absence of chemical reactions (i.e., in the absence of annihilation or creation of particles) function $\rho(r, t)$ would have obeyed diffusion equation

$$\partial\rho(r, t)/\partial t = D\Delta\rho(r, t)$$

where D is the doubled diffusion coefficient of one particle.

Now let us assume that during time dt in arbitrary volume dV of the system $qdtdV$ particles are created and let these particles be randomly distributed in the volume. In accordance with this assumption, the right part of the diffusion equation should be complemented by a "generation" term q. Now if we assume that particles disappear due to mutual collisions (but not collisions with the selected particle) the number of particles disappearing at distance r in volume dV is proportional to particle density $\rho(r,t)$, while the proportionality factor is defined by flux $j(t)$ of probability density into one particle.

Thus, the full equation for function $\rho(r,t)$ should have the form

$$\frac{\partial \rho(r,t)}{\partial t} = D\Delta \rho(r,t) + q - j\rho(r,t) \tag{3.9}$$

where

$$j(t) = 4\pi DR^2 \left.\frac{\partial \rho}{\partial r}\right|_{r=R}$$

(R is the doubled radius of the particles).

Let us note that

$$\lim_{r\to\infty} \rho(r,t) = \rho_0(t)$$

where $\rho_0(t)$ is the average particle concentration.

According to the Smoluchowski method, we must determine flux j into one particle from the diffusion equation in the case of random initial distribution and then solve equation

$$\partial \rho_0(t)/\partial t = -j\rho_0(t) + q \tag{3.10}$$

which evidently cannot be reduced to Equation (3.9). Equation (3.9) has a stringent base in the case of small particle density, when $\rho_0 R^3 \ll 1$.

Sometimes $\rho_0 R^3 \sim 1$, for example, with polymerization at deep transformation degrees of monomers. Then in the course of chemical transformation the viscosity in the system changes, and consequently the diffusion coefficient. These cases require special analysis.

3.2. METHOD OF INTERVALS

The Method of Intervals based on specially built two-particle distribution functions [147,148] is one of the ways to take into account particle correlation.

Let us proceed from the following assumptions: each active center (particle) performs the random walk with an effective diffusion coefficient; upon each particle encounter a reaction occurs with a unit probability, i.e., each particle acts as an absolute absorbing screen for any other particle; particle concentration in the system is low, i.e., the probability of a three- or more particle collision is ignored.

Let us start by investigating the kinetics of the simplest reactions of type

$$A + A \longrightarrow B \tag{3.11}$$

which play an appreciable role in radiational-chemical transformations in solids.

Before introducing the necessary binary correlation function, let us first consider the discrete in time and space process of random particle walks. Let the system be a planar lattice with interknot spacing a. The points of the lattice are occupied by particles. In time τ all particles move to a step in one of the four possible directions.

Let

$$am_{ij} = (x_i - x_j) \qquad an_{ij} = (y_i - y_j)$$

where (x_i, y_i) are the coordinates of an i-particle.

Now we introduce function $N(m, n, t)$ which will determine the number of radius-vectors between arbitrarily chosen particles having projections m and n, i.e.,

$$N(m, n, t) = \sum_{j>i} \delta(m - m_{ij})\delta(n - n_{ij}) \tag{3.12}$$

where $\delta(m - m_{ij})$ is Kronecker's symbol.

The normalization condition for function $N(m, n, t)$ reads as

$$\sum_m \sum_n N(m, n, t) = \frac{1}{2} M(M - 1) = N$$

where $M(t)$ is the full number of particles in the system at time t.

It can be shown [148] that in the absence of collisions, function $N(m, n, t)$ satisfies the usual diffusion equation.

Let Equation (3.12) be valid at time t. The displacement of an i-particle in time τ shall be denoted by unit vector ξ_i, which can take four possible values: (0,1); (0,−1); (1,0); and (−1,0). Consequently, during step τ there are 4^M versions to change the system. It is evident that

$$N(m, n, t + \nu) = \sum_{j>i} \delta[m - (m_{ij} + \xi_x{}^i - \xi_x{}^i)]\delta[n - (n_{ij} + \xi_y{}^i - \xi_y{}^j)]$$

By $S_{\vec{\xi}}[f(\vec{\xi}_1, \vec{\xi}_2, \ldots, \vec{\xi}_m)]$, we shall denote the operation of averaging the function $f(\vec{\xi}_1, \ldots, \vec{\xi}_m)$ over all the possible displacement vectors $\vec{\xi}_i$.

The averaging of function $N(m,n,t+\tau)$ over $\vec{\xi}$ gives (at $V = 1$):

$$S_{\vec{\xi}}N(m,n,t+\tau) = S_{\vec{\xi}} \sum_{j>i} \delta(m - m_{ij} - \xi_x^i + \xi_x^j)\,\delta(n - n_{ij} - \xi_y^i + \xi_y^j) = \frac{1}{16}\,[4N(m,n,t) + 2N(m+1,n+1,t) + 2N(m+1,n-1,t) + 2N(m-1,n+1,t) + 2N(m-1,n-1,t) + N(m+2,n,t) + N(m-2,n,t) + N(m,n+2,t) + N(m,n-2,t)] = F(m,n,t)$$

At the same time, it can be easily shown that the dispersion of value $N(m,n,t+\tau)$ tends to zero when the volume of system V and the number of particles M tend to infinity, so that $\rho_0 = M/V$ remains a finite value [148]:

$$S_{\vec{\xi}}N^2(m,n,t+\tau) - [S_{\vec{\xi}}N(m,n,t+\tau)]^2 = S_{\vec{\xi}}\Bigg[\sum_{j>i}\sum_{j'>i'} \delta(m - m_{ij} - \xi_x^i + \xi_x^j)\,\delta(n - n_{ij} - \xi_y^i + \xi_y^j) \times \delta(m - m_{i'j'} - \xi_x^{i'} + \xi_x^{j'})\,\delta(n - n_{i'j'} - \xi_y^{i'} + \xi_y^{j'})\Bigg] - [S_{\vec{\xi}}N(m,n,t+\tau)]^2 \longrightarrow 0$$

Hence with this limit

$$N(m,n,t+\tau) = S_{\xi}N(m,n,t+\tau) = F(m,n,t)$$

But this equality is equivalent to the diffusion equation for $N(m,n,t)$, i.e., on average the system of intervals behaves as a system of independent intervals.

Now let us take into consideration the fact that particle collisions occur in the system which result in their disappearance. To find the collision term in our equations, let us determine the full number and the distribution of intervals disappearing upon collision of particle pairs. We shall return to the discrete model. Let $2A$ particles disappear in time τ

$$A = \frac{1}{16}\,[N(+2,0) + N(0,+2) + 2N(+1,-1)]^*$$

---

*Hereinafter in the text $N(m,n,t)$ means values averaged over ξ_i in the above mentioned sense, i.e. the corresponding dispersions are small.

Each of these intervals connected a separate pair of particles and, hence, among these A intervals which are missing, are those which have a common particle (point). Value A contains no information on the competition of particles. It is connected with the conditional probability of disappearance of a given number of intervals on the condition that we know exactly which intervals left the system. Thus, consideration of particle competition is equivalent to finding the absolute probability of disappearance of the A intervals.

In the system of connected intervals discussed here, upon the disappearance of $2A$ particles, the number

$$C_2^{2A} \simeq 2A^2$$

of intervals connecting these particles quit. In addition, all the intervals connecting the remaining particles with the ones leaving the system also disappear; their number is $2A(M - 2A)$. Now the total number of particles disappearing during time τ is

$$2A^2 + 2A(M - 2A) \simeq 2AM$$

Here we consider $A \ll M$, which is valid in the case of low density.

The distribution function of the disappearing intervals has the form $2AMN(m_x, m_y, t)/N$. This consideration is valid if M and V tend to infinity, and it can be proved analogous to the statement on the independence, on average, of intervals in a system.

Now the full kinetic equation can be written

$$N(m_x, m_y, t + \tau) = F(m_x, m_y, t) - 2AM\frac{N(m_x, m_y, t)}{N} \tag{3.13}$$

If we proceed to continuous variables $x = am_x$ and $y = am_y$ we obtain from Equation (3.13):

$$\frac{\partial N(x, y, t)}{\partial t} = D_A N(x, y, t) - 2\bar{A}M\frac{N(x, y, t)}{N} \tag{3.14}$$

where $D = a^2/\tau$, $\bar{A} = \lim\limits_{a\to 0, \tau\to 0} A$

The presence of absolutely absorbing screens is equivalent to the boundary condition on circle $(x^2 + y^2)^{1/2} = R$ ("two-dimensional" particles are chosen in the form of circles of radius R):

$$N(x, y, t)\Big|_{(x^2 + y^2)^{1/2} = R} = 0 \tag{3.15}$$

Note that the normalization condition in the general three-dimensional case assumes the form:

$$\int N(x, y, z, t)dV = \frac{1}{2}M^2V$$

Defining value A as the number of intervals which disappeared in the system of independent intervals under condition (3.15), one must assign the value with a flux through the hypersurface, depending on the particle form. Then

$$\bar{A} = D \int \nabla N(x, y, z, t) d\vec{S}$$

In further discussion, it is convenient to deal with a new function

$$\rho(\vec{r}, t) = 2N(\vec{r}, t)/MV \tag{3.16}$$

instead of function $N(r, t)$.

Function ρ signifies the particle density separated by distance $r = (x^2 + y^2 + z^2)^{1/2}$ and it satisfies the normalization condition

$$\int \rho(\vec{r}, t) d\vec{r} = M(t)$$

Let us illustrate its introduction for the case of cylindric symmetry. Under condition (3.16) it follows from Equation (3.14):

$$\frac{1}{2}\left[M \frac{\partial \rho}{\partial t} + \rho \frac{dM}{dt}\right] = \frac{1}{2} MD\Delta\rho - M2\pi DR \left.\frac{\partial \rho}{\partial r}\right|_{r=R} \rho \tag{3.17}$$

where

$$\bar{A} = 2\pi DR \left.\frac{\partial \rho}{\partial r}\right|_{r=R}$$

Integrating over the entire space one obtains

$$\frac{dM}{dt} = 2\pi DRM \left.\frac{\partial \rho}{\partial r}\right|_{r=R} \tag{3.18}$$

Substitution of (3.18) into (3.17) gives the equation

$$\frac{\partial \rho}{\partial t} = D\Delta\rho - 2\pi DR \left.\frac{\partial \rho}{\partial r}\right|_{r=R}$$

In the three-dimensional case, at spheric symmetry of the problem, one obtains

$$\frac{\partial \rho}{\partial t} = D\Delta\rho - \alpha^2(t)\rho(\vec{r}, t) \tag{3.19}$$

where

$$\alpha^2(t) = 4\pi DR^2 \left.\frac{\partial \rho}{\partial r}\right|_{r=R} = j(t) \tag{3.20}$$

Equation (3.19) is the basic one in this theory. Such equations were first considered by Antonov-Romanovsky [149], though his distribution function had an absolutely different physical meaning.

The boundary conditions used are the consequence of the inherent notion in the model of a particle as an absolutely absorbing screen which is equivalent to the Smoluchowski boundary condition

$$\rho(R, t) = 0 \tag{3.21}$$

Generally, this condition can be replaced by an expression suggested by Collins and Kimball [18]*.

The initial conditions are chosen depending on the concrete physical situation. As an example, let us consider the simplest situation when the initial particle distribution is random. This was the situation analyzed by Smoluchowski. So let

$$\rho(r,0) = \rho_0 \tag{3.22}$$

To solve (3.19) under condition (3.21), we shall introduce function $W(r,t)$ defined by the relationship

$$\rho(r, t) = r^{-1} W(r, t) \exp\left(- \int_0^t \alpha^2 d\tau \right)$$

As a result we obtain a simple diffusion equation for $W(r,t)$ with the following initial and boundary conditions: $W(R,t) = 0$, $W(r,0) = \rho_0 r$.

The solution of such an equation is readily calculated. Let us write the result for function $\rho(r,t)$:

$$\rho(r, t) = \rho_0 r^{-1} \left[r - R + R\Phi\left(\frac{r - R}{\sqrt{4Dt}} \right) \right] \exp\left(- \int_0^t \alpha^2 d\tau \right)$$

where

$$\Phi(x) = \frac{2}{\sqrt{\pi}} \int_0^x d\xi \exp(-\xi^2)$$

---

*The derivation and discussion of the general expressions for boundary conditions will be given in Section 4.2.

Now according to (3.20), we find the flux through the absorbing sphere–the sphere of radius R:

$$\alpha^2(t) = 4\pi DR\rho_0[1 + R(\pi Dt)^{-1/2}] \exp\left(-\int_0^t \alpha^2 d\tau\right) \tag{3.23}$$

Note that integrating (3.19) over the whole volume one obtains

$$dM/dt = -\alpha^2 M \tag{3.24}$$

At $M(0) = M_0$ we get the solution of (3.24) in the form

$$M(t) = M_0 \exp\left(-\int_0^t \alpha^2 d\tau\right) \tag{3.25}$$

Simple transformations of Equation (3.23) yield an equation for $\alpha(t)$:

$$-\alpha^2 = R\cdot\alpha^{-2}\cdot 4\pi D\rho_0\left(1 + \frac{R}{\sqrt{\pi Dt}}\right)\frac{d}{dt}\left\{\alpha^2\left[4\pi D\rho_0 R\left(1 + \frac{R}{\sqrt{\pi Dt}}\right)\right]^{-1}\right\} \tag{3.26}$$

Let

$$f = \alpha^2[4\pi D\rho_0 R(1 + R/\sqrt{\pi Dt})]^{-1}$$

The following equation is readily obtained from (3.26)

$$\frac{df}{dt} = -4\pi D\rho_0 R(1 + R/\sqrt{\pi Dt})f^2 \tag{3.27}$$

Value f is some dimensionless value having the sense of the reduced concentration. Then Equation (3.27) has the meaning of a kinetic equation of a bimolecular reaction with a time dependent rate constant.

At large time, the second term in the brackets of Equation (3.27) disappears, i.e., within this limit the value of the rate constant decreases to some finite value. As could be expected, we obtain a simple chemical kinetic equation.

Solving (3.27) and (3.25) jointly gives

$$M(t) = M_0\left[4\pi R\rho_0\left(\frac{2R}{\sqrt{\pi}}\sqrt{Dt} + Dt\right) + 1\right]^{-1} \tag{3.28}$$

This result coincides exactly with the corresponding result of Smoluchowski [26].

Now we shall derive the general equation for a source of particles. Let us consider the general case when the particles created are correlated and are created, for example, in bunches. We shall proceed from the discrete model of the process. Let us assume that in time τ the number ΔM_b of particles is created in bunches, the number of particles in a bunch being constant and equal to N_B. The probability of the intervals of r-length emerging in a bunch is defined by function

$$dV V_b^{-1} \exp(-r/r_b) \tag{3.29}$$

where V_B is the volume of the bunch; r_b is the effective size of the bunch.

Besides the known term of particle disappearance, Equation (3.13) for function $N(m, n, t)$ should include additional terms, which can be determined from the discussions that follow.

During time step τ intervals appear between particles that were in the system earlier as well as ones from new bunches. The number of new intervals is

$$M \Delta M_b \tag{3.20}$$

while the distribution in lengths is equiprobable, since the gravity centers of bunches are randomly arranged.

Besides, intervals appear connecting particles inside the bunches. Their full number is

$$\frac{1}{2} N_b{}^2 N_1 = \frac{1}{2} \Delta M_b N_b \tag{3.31}$$

where N_1 is the number of the emerging bunches: $N_1 = \Delta M_b / N_b$.

Taking into account (3.29), the number of intervals of length equals

$$\frac{1}{2} \Delta M_b N_b \exp(-r/r_b) dV/V_b$$

Now let us take into consideration the intervals connecting the particles of separate bunches. The number of these intervals of r-length is equal to

$$\frac{1}{2} (\Delta M_b)^2 dV/V$$

However, in further discussion this term will be omitted since $\Delta M_b \ll$ $\ll M_b$. Now proceeding to continuous variables and assuming that $\tau^{-1} \Delta M_b = VI$ (where I is the intensity of particle creation), $M/V = \rho_s$ is the average concentration of particles in the system, $N_b/V_b = \rho_b$ is the concentration of particles in a bunch, one can write the equation for function $N(x, y, t)$. We shall consider the case of spheric symmetry.

Then

$$\frac{\partial N(r,t)}{\partial t} = D\Delta N - 8\pi DR^2 \frac{\partial N}{\partial r}\bigg|_{r=R} M \frac{N(r,t)}{N} + VI\left[\rho_s + \frac{1}{2}\rho_b \exp(-r/r_b)\right] \tag{3.32}$$

and

$$\overline{A} = 4\pi DR^2 \frac{\partial N}{\partial r}\bigg|_{r=R}$$

As before, let us proceed to function $\rho(r,t)$, defined by relationship (3.116). Substituting (3.16) into (3.32) and integrating the deduced expression with respect to volume gives

$$2\frac{dM}{dt} = -8\pi DR^2 \frac{\partial \rho}{\partial r}\bigg|_{r=R} M + 2IV \tag{3.33}$$

Let us introduce function α_1^2:

$$\alpha_1^2(t) = \alpha^2(t) + I\rho_s^{-1} \tag{3.34}$$

Now from (3.32–3.34) one can derive the final form of equation

$$\frac{\partial \rho(r,t)}{\partial t} = D\Delta\rho - \alpha_1^2\rho(r,t) + 2I\left[1 + \frac{1}{2}\rho_s^{-1}\rho_b \exp(-r/r_b)\right] \tag{3.35}$$

which is, in fact, the general equation needed with a particle source. Note that radiation intensity I appears not only in the particle "creation" term, but also in the "reaction" term α_1^2, which is accounted for by the fact that correlation in the particle position is taken into account.

It should be noted that following the generally accepted method of analysis, we would have proceeded from an equation of type (3.10) having added to its right part a term proportional to intensity I of particle creation, but then the reaction term would be proportional to α^2, not to α_1^2, and it would have been independent of I.

3.2.1. Application of the Method of Intervals to One-Dimensional Problems

Using the Method of Intervals, let us consider one-dimensional problems which, as will be seen below, have their own specificity. Such problems are important for studying the kinetics of radical recombination in polymers [150], the recombination of defects in solids [151], and transport through selective channels of biological membranes [152–153]. In particular, the migration of free valency in long linear molecules (like polyethylene or rubber) may carry a one-dimensional character [18].

Disregarding the mechanism of migration, we can consider that an active particle (henceforth the term for the migrating free valency) moves in a medium characterized by a diffusion coefficient. The act of recombination will then represent the encounter of two active particles and their "sticking together." The specificity of the general recombination process is determined by the fact that only neighboring particles can react, whereas the nonneighboring ones can enter into reaction only when the particles that are between them disappear (such a situation in Section 1.2 is called the "trolley-bus" effect).

Our interest lies with determining the analytic time dependence of particle concentration $N(t)$ in a polymer chain [154].

Let us consider a one-dimensional problem, i.e., assume that the polymer chain is an axis of length L consisting of links, the length of bond C—C being recognized at link (d). N particles with coordinates x_i are located on the axis:

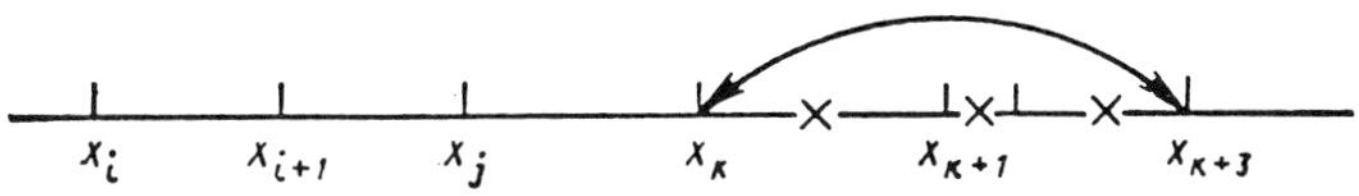

Let us assume that each particle in some characteristic time τ can displace with equal probability either to the left or to the right by one link of length d. This walk is a diffusion process and it can be discussed using a diffusion equation ($D = d^2/2\tau$). We shall assume that two particles upon collision quit the system. Therefore, the place of the collision can be occupied by other particles. Note that in the general case the model can be made more complicated to describe other possible systems (with partial penetrability of the collision site with a "sticking together" probability smaller than one, etc.).

Finding the exact solution of the posed problem appears complicated though apparently possible. Hence, we shall first deduce an equation approximately describing the given system. Let us consider a process discrete both with respect to variable N and the time variable.

Let the system at time t assume the form shown above. If during the next time interval τ, distance $(0,X_i)$ increases by a unit, then distance (X_i, X_{i+I}) may remain unchanged in the case of a step of X_{i+I}, particle to the right and may decrease by two units in the case of a step of X_{i+I} to the left. Distance (X_{i+I}, X_{i+2}) may increase by two units or preserve its value if (X_i, X_{i+I}) has decreased; and it may decrease by two units or preserve its value if (X_i, X_{i+I}) has remained unchanged.

Let us designate the increase of the distance between neighboring particles by sign "+" in a circle, and decrease by sign "−", the preservation of the interval by a hollow circle. Then the above discussion may be presented in the form of diagram:

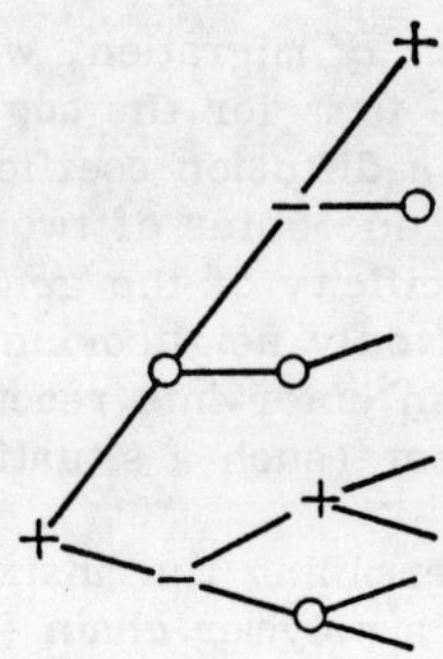

It should be stressed that following each "−" in each branch is necessarily a hollow circle, or "+". Consequently, of all the intervals about half will preserve their length, a quarter—will increase their length by two units and about a quarter will decrease it.

Now it is more convenient to introduce a new discrete variable in the distance between the neighboring particles and in further discussion to deal with distribution function $N(m,t)$—the number of neighboring distances of value m at time t. Then the full number of particles on the straight line will be described by expression

$$\sum_{m} N(m,t) = N(t) \tag{3.36}$$

Besides, on average, the number of "+" is equal to the number of "−" and half the number of hollow circles, the sum of which is $N/4$. Now one can write equation

$$N(m,t+\tau) = \frac{1}{2}N(m,t) + \frac{1}{4}N(m-2,t) + \frac{1}{4}N(m+2,t) \tag{3.37}$$

In time τ the number $N(2,t)/4$ of intervals disappears, so

$$\sum_{m=2} N(m,t+\tau) = \sum_{m=2} N(m,t) - \frac{1}{4}N(2,t)$$

Here we assume that $N(0,t) = 0$ which corresponds to the absorption.

However, Equation (3.37) does not fully describe the situation since, together with the disappearing interval, the two neighboring ones disappear but a "triple" one appears. This is taken into account in the following way. The contribution of disappearing intervals relative to total number is

$$2N(2,t)\left\{4\left[\sum_{m} N(m,t) - \frac{1}{4}N(2,t)\right]\right\}^{-1}$$

At the same time 1/4 $N(2,t)$ intervals appear with a new distribution function $F(m)$ which can be written in the form

$$\tilde{F}(m) = \sum_{m'} \frac{F(m')F(m - m')}{\left[\sum_m N(m,t) - \frac{1}{4} N(2,t)\right]^2}$$

where

$$F(m') = \frac{1}{2} N(m',t) + \frac{1}{4} N(m' - 2,t) + \frac{1}{4} N(m' + 2,t)$$

Expression $F(m')/[\Sigma_m N(m,t) - 1/4\ N(2,t)]$ signifies the probability of the presence of intervals of length m' at time t.

With this correction the full equation takes the form:

$$N(m,t + \tau) = F(m)\left\{1 - \frac{1}{2} N(2,t)\left[\sum_m N(m,t) - \frac{1}{4} N(2,t)\right]^{-1}\right\} + \\ + \frac{1}{4} N(2,t)\tilde{F}(m) \tag{3.38}$$

It is easily seen that condition (3.36) is satisfied. The continuous analog of Equation (3.38) has the form:

$$\frac{\partial f(\xi,t)}{\partial t} = D \frac{\partial^2 f}{\partial \xi^2} + D \left.\frac{\partial f}{\partial \xi}\right|_{\xi=0} \cdot \left\{\left[\int_0^\infty f(\xi,t)d\xi\right]^{-2} \int_0^\xi f(\xi',t)f(\xi - \xi',t)d\xi' - 2f(\xi,t) \times \right. \\ \left. \times \left[\int_0^\infty f(\xi,t)d\xi\right]^{-1}\right\} \tag{3.39}$$

with boundary conditions

$$f(0,t) = 0 \qquad f(\infty,t) = 0 \tag{3.40}$$

and initial condition

$$f(\xi,0) = N_0 \delta e^{-\delta\xi}, \qquad \delta = N_0/L \tag{3.41}$$

[The type of function $f(\xi,0)$ is chosen on the basis of the assumption that at the initial time moment the particles are distributed randomly on the straight line]. The obtained equation is complicated, thus we shall investigate two limit cases—of small and large times. Equation (3.39) is solved using the half Fourier transform. Omitting fairly long calculations, we obtain for small t:

$$N(t) = N_0\left(1 - 4\sqrt{\frac{Dt}{\pi}}\ \delta\right) \tag{3.42}$$

where $\delta = N_0/L$ is the initial concentration of particles on the straight line.

And for large t:

$$N(t) = (2DtL^{-2}\ N_0^{-2})^{-1/2} \sim t^{-1/2}$$

Result (3.42) has been used [154] for analyzing the kinetics of radical recombination in polyethylene; a satisfactory agreement of theoretical and experimental dependences was obtained. This agreement can be more complete if partial penetrability of the sites of particle collisions is taken into account and a more realistic initial condition is used, i.e., considering the possibility of a bunch initial particle distribution.

In conclusion, let us consider the phenomenon of limit concentration of radicals arising upon their accumulation in solids.

Let in time interval $(t, t + \Delta t)$ appear $I_0\Delta t$ particles on the straight line. The conditions of particle disappearance remain the same. We shall presume that the particle created on interval ξ annihilates this interval and two new intervals ξ and $\xi' - \xi$ emerge. Let $\varphi(\xi,\xi')$ be the probability density of the appearance of interval ξ from ξ', and

$$\int_0^{\xi'} \varphi(\xi,\xi')\ d\xi = 1$$

Assuming that the emergence of an interval of any length smaller than ξ is equiprobable one obtains

$$\varphi(\xi,\xi') = \begin{cases} 1/\xi' & \text{if} \quad \xi < \xi' \\ 0 & \text{if} \quad \xi > \xi' \end{cases}$$

The probability of creating particles to be in interval ξ' equals $L^{-1}\xi' f(\xi') = W(\xi')$. Then the absolute probability of the appearance of interval ξ is

$$2\int_{\xi}^{\infty} d\xi' L^{-1} f(\xi') = 2\int_{\xi}^{\infty} W(\xi')\,\varphi(\xi')\,d\xi'$$

The multiplier in front of the integral is stipulated by the probability for the emergence ξ in any of the two intervals formed upon the "division" of interval ξ'. In time interval $(t, t + \Delta t)$ the number $2(I_0/L)\Delta t \int_\xi^\infty f(\xi')d\xi'$ of intervals emerge and the number $I_0 L^{-1} \xi f(\xi)\Delta t$ of intervals disappear. Hence, the total equation can be written in the form

$$\frac{\partial f}{\partial t} = \frac{I_0}{L}\,[2N(\xi) - \xi f(\xi)] + D\left\{\frac{\partial^2 f}{\partial \xi^2} + \right.$$

$$\left. + x\left[N^{-2}(0)\int_0^\xi f(\xi')f(\xi - \xi')d\xi' - 2N^{-1}(0)f(\xi)\right]\right\}$$

where

$$x = \left.\frac{\partial f}{\partial \xi}\right|_{\xi=0}, \qquad N(\xi) = \int_\xi^\infty f(\xi')d\xi'$$

with boundary conditions (3.40) and initial condition (3.41). It is easily seen that the deduced equation has a stationary solution, although it is quite difficult to find it. The dependence of limit concentration on radiation intensity I_0 can be estimated on the basis of dimensionality:

$$\delta_{\text{lim}} = N_{\text{lim}}/L \sim \sqrt[3]{I_0/(LD)}$$

3.2.2. Kinetics of Diffusion Controlled Processes on a Surface

It is well known that taking into account transport stages of approaching reagents leads to a time dependence of the effective rate constant k_{ex}. For three-dimensional systems at large time $t \gg R^2/D$ (where R is the size of particles, D is the diffusion coefficient) rate k_{ex} becomes constant, and the equations of diffusion kinetics lead to the same kinetic laws as the equations of formal kinetics. However, in the case of two-dimensional systems, the situation may appear to be quite different. In particular, in a bimolecular process on the surface at large time ($Dt/R^2 \gg 1$, $Dt \gtrsim > (nR^2)^{-1}$ where n is the mean surface concentration of reagents) concentration ρ decreases

$$\rho \sim t^{-1} \ln (Dt/R^2)$$

whereas formal kinetics predicts dependence $\rho \sim t^{-1}$. This means that k_{ex} is weakly (logarithmically) time dependent.

Let us consider the kinetics of diffusion controlled reactions on the surface at random arrangement of interacting particles. In the general case it is necessary to take into consideration the influence of initial fluctuations in the location of reagents on the surface which as it is known (see [155] and section 3.6) can considerably affect the asymptotic behavior of kinetic dependences at large times.

For solving this problem, it appears convenient to apply the Method of Intervals. This method provides for describing the kinetics of process $C + A \to CA$ occuring between diffusing particles C and A by analyzing diffusion equations in the space of intervals:

$$\partial \rho_C/\partial t = D_1 \Delta \rho_C - \beta^2 \rho_C(r,t) \tag{3.43}$$

$$\partial \rho_A/\partial t = D_2 \Delta \rho_A - \beta^2 \rho_A(r,t) \tag{3.44}$$

Here

$$\rho_C = \frac{2N_{CC}(r,t)}{SM_C(t)}, \qquad \rho_A = \frac{N_{CA}(r,t)}{SM_C(t)}, \qquad \beta^2 = 2\pi DR \left.\frac{\partial \rho_A}{\partial r}\right|_{r=R} \tag{3.45}$$

where $N_{CC}(r,t)$ and $N_{CA}(r,t)$ are the distribution functions between particles of one type and particles of various types respectively; $M_C(t)$ and $M_A(t)$ are the corresponding numbers of particles of type C and type A in a circle of area S; R is the "radius" of the reaction.

Functions $\rho_C(r,t)$ and $\rho_A(r,t)$ are normalized on the total number of particles in this circle:

$$\int \rho_C(r,t)dS = M_C, \qquad \int \rho_A(r,t)dS = M_A$$

so the integration of Equation (3.43) and (3.44) over the surface yields

$$dM_C/dt = dM_A/dt = -\beta^2 M_C \tag{3.46}$$

If in the three-dimensional case $\beta^2 = kM_A$, i.e., Equation (3.46) is reduced to a usual (in formal chemical kinetics) equation for a bimolecular reaction, then in our two-dimensional case (the problem of diffusion kinetics on a surface), value k appears to be a time dependent function.

Let us consider the simplest case, when all particles in the system are the same ($C = A, D_1 = D_2 = D$) and they enter into reaction with each other (the bimolecular process). In this case, it is possible to confine ourselves to Equation (3.43). (Index "C" will be omitted in further discussion). Let us find its solution under initial condition $\rho(r,0) = c_0$ and the simplest boundary condition

$$\rho(R.t) = 0 \tag{3.47}$$

Substitution

$$\rho(r,t) = W(r,t) \exp\left(-\int_0^t \beta^2 d\tau\right)$$

leads to a diffusion equation for function $W(r,t)$ under conditions $W(r,0) = c_0$ and (3.47). The solution of such an equation is known and has the form:

$$\rho(r,t) = \frac{2c_0}{\pi} \exp\left(-\int_0^t \beta^2 d\tau\right) \int_0^\infty \exp(-Du^2 t)\, du \times \frac{[J_0(uR)N_0(ur) - J_0(ur)N_0(uR)]}{u[J_0^2(uR) + N_0^2(uR)]} \tag{3.48}$$

where $J_0(z)$ and $N_0(z)$ are Bessel and Neumann functions of zero order respectively.

From (3.48) the self-consistent condition follows:

$$\beta^2 = \frac{8DC_0}{\pi} \exp\left(-\int_0^t \beta^2 d\tau\right) \int_0^\infty \frac{du \exp(-Du^2 t)}{u[J_0^2(uR) + N_0^2(uR)]} \tag{3.49}$$

From (3.46), (3.48) and (3.49) one obtains solution

$$M(t)M_0^{-1} = \left\{ \int_0^t d\tau \frac{8DC_0}{\pi} \int_0^\infty du \exp(-Du^2 t) u^{-1} \times [J_0^2(uR) + N_0^2(uR)]^{-1} + 1 \right\}^{-1}$$

which at large times yields an asymptotic dependence

$$M(t) = M_0(\rho_0 \cdot 4\pi Dt)^{-1} \ln(4Dt/R^2) \tag{3.50}$$

The possibility of a non-stationary course of the radical recombination reaction "from the beginning and from the end" (because of the change in dimensionality of the process, i.e., for one-dimensional and two-dimensional processes) is considered in [150]. For the two dimensional case it has been determined that $M(t) \sim (\ln t)^{-1}$. Comparison to (3.50) shows that the consideration of the correlation between particles changes the asymptotic behavior of $M(t)$. Experimental dependences $M(t) \sim (\ln t)^{-1}$ are discussed in [156,157].

3.2.3. Kinetics of Diffusion Controlled Recombination of Radicals

3.2.3.1. Bunch Particle Distribution Model

The simplest reaction in diffusion kinetics is the one described by the scheme

$$A + A \longrightarrow B \tag{3.51}$$

and occurring between active centers of one type not interacting at large distances. An example of a reaction of this type is the recombination of radicals emerging in condensed media upon exposure to penetrating radiation [18,101,158–161]. As has been established [159], the interaction between radicals separated by distance > 1 nm can be neglected. In this case, an investigation of the kinetics of radical recombination permits analyzing the influence of correlation between particles upon DCR kinetics. This possibility is determined by the fact that radicals are, as a rule, formed in tracks where their local concentration ρ_b is much higher than the mean over volume concentration ρ_s. Thus it is possible to speak about the inhomonenic initial distribution of active centers and to pose a problem about the effect of initial particle distribution upon the character of the kinetic dependences of reactions (3.51). Such a problem is considered in [147] within the framework of the Method of Intervals; here are some of the results of this work.

Let us first consider a three-dimensional problem (assuming the spheric symmetry of the system); and write the principal equation of the Method of Intervals:

$$\partial \rho(r,t)/\partial t = D\Delta\rho(r,t) - \alpha^2(t)\rho(r,t) \tag{3.52}$$

where

$$\alpha^2(t) = 4\pi D R^2 \left.\frac{\partial \rho(r,t)}{\partial r}\right|_{r=R} \tag{3.53}$$

which we shall solve using the Smoluchowski boundary condition $\rho(R, t) = 0$. To evaluate the influence of the above-mentioned inhomogeneity in the initial radical distribution upon the character of the kinetic dependences, we take as the initial condition distribution*:

$$\rho(r, 0) = \rho_s + \beta(r)\rho_b \tag{3.54}$$

function $\beta(r)$ being such that at large distance $\rho(r,0) \to \rho_s$, while at small distances $\rho(r,0) \to \rho_b$.

It is assumed for simplicity that

$$\beta(r) = \begin{cases} 1 & \text{if} \quad r < r_b \\ 0 & \text{if} \quad r > r_b \end{cases}$$

where r_b is the effective size of the bunch.

---

*More "realistic" functions characterizing the initial distribution are discussed in reviews [101,159,162].

The solution of Equation (3.52) with condition (3.54) is rather simple, but it leads to a bulky expression for the full concentration of particles $M(t)$.

Its detailed derivation and the expression of the function are given in [147]. Here only the asymptotics of the function will be presented.

At short times, when $r_b^2(Dt)^{-1} \gg 1$ one obtains

$$M(t) = M_0\left\{4\pi\rho_s R\left[2R\sqrt{\frac{Dt}{\pi}}\,(1 + \rho_b\rho_s^{-1})\right] + 1\right\}^{-1}$$

i.e., the kinetics are determined by the concentration of radicals inside the bunch:

$$M(t) \simeq M_0(8\sqrt{\pi Dt}\;R^2\rho_b + 1)^{-1} \tag{3.55}$$

At large times when $Dt \gg r_b^2$

$$M(t) = M_0(4\pi DtR\rho_s + 1)^{-1} \tag{3.56}$$

This expression coincides with the result of the usual Smoluchowski method and shows that within this limit the kinetics are defined by general concentration ρ_s.

Dependences of type (3.55) have more than once been investigated experimentally; they are discussed in detail in review [114]. It follows from the results obtained in [114] that the effective constant of reaction rate k_{ex} has a considerably more complicated form than in traditional theory, where $k_{ex} = 4\pi DR[1 + R(\pi Dt)^{-1/2}]$. This may be connected with the inhomogeneity of the initial radical distribution. Indeed, if the generally accepted theory is applied, then in analyzing the experimental results using "root" anamorphisms the effective rate constant $\tilde{k}$ is defined by expression [114]:

$$\tilde{k} = 8R^2\sqrt{\pi D}\;\rho_s \tag{3.57}$$

If the bunch character of initial particle distribution is taken into account, result (3.57) is valid only at large times. At short times, value ρ_s in (3.57) should be substituted for value ρ_b—the particle concentration in a bunch: $\tilde{k}_b = 8R\sqrt{\pi D}\;\rho_b$. In particular, this shows that at $\rho_b \gg \rho_s$ only $\tilde{k}_b$ can be practically determined. In the general case $\tilde{k}$ depends both on ρ_s and on ρ_b which makes the general expression for k_{ex} more complicated.

If values ρ_s and ρ_b are prescribed, it is possible to evaluate the time after which the recombination of particles belonging to different bunches begins to compete with that of particles inside the bunch. For this time T one obtains:

$$\rho_s DT \simeq \frac{2}{\sqrt{\pi}}\;\rho_b\;\frac{r_b}{4}\;\sqrt{4DT}$$

$$T \simeq \frac{r_b^2}{\pi} \frac{\rho_b^2}{\rho_s^2 D}$$

3.2.3.2. Dependence of Limit Concentrations on Radiation Intensity. Models With Random and Bunch Distribution

The correlation between the particles created by irradiation influences not only the kinetics of thermal particle annealing (following the cessation of exposure to radiation) but also the process of radical accumulation upon exposure to penetrating radiation. Viewed this way, establishing the dependences of limit radical concentrations ρ_s on radiation intensity I at different distribution of the created particles would be of interest. As has been shown in monograph [18], complex dependences between ρ_s and I are possible. These dependences can be linear $\rho_s \sim I$ or "root" $\rho_s \sim I^{1/2}$; ρ_s may also be independent of I.

Let us discuss the possible reasons of such dependence of ρ_s on I proceeding from the different distributions of the created active centers. First we shall consider the regularities of $\rho_s(I)$ at random particle distribution assuming that $\rho_b = 0$. Since we are interested in the stationary case, the problem is reduced to solving a linear inhomogeneous equation (the case of spheric symmetry)

$$D(1/r)(\partial^2 \rho r/\partial r^2) - \alpha_1^2 \rho + 2I = 0$$

under conditions $\rho(R) = 0$, $\rho(\infty) = \rho_s$. The corresponding solution has the form:

$$\rho(r) = \frac{2I}{\alpha_1^2}\left\{1 - \frac{R}{r} \exp\left[-\frac{\alpha_1}{\sqrt{D}}(r - R)\right]\right\} \tag{3.58}$$

Value $\alpha_1/\sqrt{D}$ has the dimensionality of reverse length and characterizes the rate of growth of function $\rho(r)$ from zero at $r = R$ to $\rho_s = 2I/\alpha_1^2$.

It can be considered that at distances $r > \sqrt{D}/\alpha_1$ distribution function $\rho(r)$ becomes practically constant while the contribution of intervals $r < \sqrt{D}/\alpha_1$ is small. Note that the expression obtained for $\rho(r)$ differs from the corresponding Smoluchowski function which has the form $\rho(r) = \rho_0(1 - R/r)$ and increases much slower with the increase of r. It follows from (3.58) that

$$\rho_\infty = \rho_s = 2I/\alpha_1^2 \tag{3.59}$$

Now we obtain the expression for α^2. From (3.58) we derive:

$$\alpha^2 = 4\pi DR\rho_s(1 + \alpha_1 R/\sqrt{D})$$

Then the dependence of limit concentration on intensity takes the form

$$\rho_s = \sqrt{I/(4\pi DR)} \tag{3.60}$$

which coincides with the corresponding result of Smoluchowski*.

However, there are certain situations when the Smoluchowski Method cannot be applied, for example when investigating the dependence of stationary concentration on intensity in the case of surface diffusion. Indeed, according to the Smoluchowski Method, we should have calculated flux $j(t)$ solving Equation (3.10) under condition (3.47). To determine the probability of a collision between two particles, $j(t)$ should have been found, but the corresponding limit equals zero, and thus the dependence of limit concentration on intensity cannot be established. The Method of Intervals serves to help find [148] this dependence:

$$\rho_s = \left[\frac{1}{2\pi D} \ln\frac{1}{R} \left(\frac{D}{8\pi I}\right)^{1/4}\right]^{1/2}$$

Now let us elucidate the influence of bunch distribution on the dependence of limit concentrations on radiation intensity. We shall once more turn to Equation (3.35) which itself provides for stationary solutions and consider the three-dimensional case. Let

$$\alpha_1 r_b/\sqrt{D} \ll 1 \tag{3.61}$$

i.e., the mean distances characterized by parameter $\sqrt{D}/\alpha_1$ are much larger than the size of bunch r_b. This approximation should correspond to the case of low particle density in a bunch. It follows from (3.35) and (3.59) that

$$\rho_s = \left[\frac{I}{4\pi DR} \left(1 - 4\pi R\rho_b r_b^{\,2}\right)\right]^{1/2} \sim I^{1/2}$$

This equation shows that approximation (3.61) is valid if condition $4\pi R\rho_b r_b^{\,2} < 1$ is satisfied, i.e., as would have been expected if particle density in a bunch is low. At $\rho_b = 0$, the result is given by Equation (3.60) which is typical for random particle distribution.

To obtain a result corresponding to the case of high particle density in a bunch, we shall make use of a "weaker" assumption $\alpha_1 R/\sqrt{D} \ll 1$, i.e., we shall presuppose the mean distances not to exceed the size of the bunch.

In this case we get the relation:

$$\rho_s = \frac{2Ir_b^{\,2}}{D\left[\sqrt{4\pi\rho_b R r_b^{\,2}} - 1\right]^2} \sim I$$

*The Smoluchowski result can be obtained from Equation (3.10).

which should be valid under condition $4\pi R \rho_b r_b^2 > 1$. Thus particle concentration in a bunch may appreciably influence the dependence of limit concentrations on radiation intensity, at low particle density in a bunch it is $\rho_s \sim I^{1/2}$ while at high particle density $\rho_s \sim I$. The expressions for ρ_s permits defining critical density $\rho_{b,cr}$ when the dependence of ρ_s on I changes:

$$\rho_{b,cr} \simeq (4\pi R r_b^2)^{-1}$$

As has been mentioned above, various cases of dependence $\rho_s(I)$ can be realized experimentally. Those interested in this problem should read monograph [18].

3.3. THE METHOD OF CELLS

The Smoluchowski problem on the diffusion of particles in the presence of absorbing traps (sinks) presents a good model for investigating the kinetics of various physical-chemical processes: the kinetics of fluorescence quenching [166], annealing of defects in irradiated materials, the growth and disappearance of pores, the segregation of a dissolved substance on different impurities [167], etc. In the theory of diffusion controlled reactions, this problem is considered for two general cases: regular arrangement of traps and chaotic (random) arrangement. In the first case, spatial symmetry considerably simplifies the solution, and this permits considering the role of competition of sinks and particle correlation. In problems with a random distribution of sinks, besides the effects of competition, it becomes most significant to consider various configurations of sinks in the general kinetics of the process. This may lead to a dependence of D on the concentration of sinks [168] (at sufficiently large times) and allows revealing the role of fluctuations in the kinetics of chemical reactions [28].

Let us consider the simplest methods of solving the diffusion problem in a medium with traps, paying particular attention to consideration of the competition of sinks (particle correlation) in DCR kinetics. We shall assume that mobile particles with concentration $\rho(\vec{r}, t)$ diffuse in a isotropic d-dimensional space where particles of another type—traps (sinks) with concentration ρ_c—are distributed in a particular way. The problem is to find the dependence of average particle concentration $\rho(t)$ or flux $j(t)$ into a trap on the time of the reaction duration. The absorbing sink is prescribed by its form and capacity, i.e., the number of particles the sink can absorb per time unit. The simplest case is the consideration of spheric or cylindric sinks of "infinite" capacity, when any number of mobile particles can be absorbed by the sink while the sink itself remains unchanged. This confirms the validity of condition $\rho_c \ll \rho(0)$. In this case, concentration $\rho(\vec{r}, t)$ is described by diffusion equation

$$\partial \rho(\vec{r}, t)/\partial t = D\nabla^2 \rho(\vec{r}, t) \tag{3.62}$$

under a definite boundary condition on the surface of the sink

$$\rho[\vec{r}(s)] = \rho_s \tag{3.63}$$

(below we shall assume ρ_s = const) and initial particle distribution:

$$\rho(\vec{r},0) = f(r) \tag{3.64}$$

The choice of the second boundary condition follows from the character of the distribution of sinks in space.

Let the sinks be arranged regularly; of interest to us is the problem of the mutual influence of sinks on the kinetics of the process. Ham [169] suggested a simple method of considering such problems—it is called the Method of Cells. At its base is the well-known mathematical procedure—expansion of the solution as a series in eigenfunctions of the corresponding boundary problem [170]. To simplify the calculations, Ham introduced an additional assumption characteristic of the Wigner-Zeitz method in analyzing the electronic energetic spectrum of solids. This form of the Method of Cells is well known [167] and found its application in radiation chemistry, in particular, in studying the diffusion of sinks in the presence of an interaction potential between the particles and the sink [84,173].

Let us first consider a cubic three-dimensional lattice formed by spherically symmetric traps with surface S and radius a. We assume that initial distribution (3.64) of the particles diffusing to sinks possess lattice symetry. Then the lattice can be broken up into equivalent symmetrical cells centered on each trap. Differential operators in (3.62) and boundary condition (3.63) do not change upon transformation, corresponding to the symmetry of the lattice, i.e., function $\rho(\vec{r},t)$ has the same symmetry as initial distribution $\rho(\vec{r},0)$. Then the problem is reduced to considering diffusion in one cell under the boundary conditions being imposed on its outer surface*. The solution of this problem can be considerably simplified if use is made of the Wigner-Zeitz method and the elementary cell of the lattice is replaced by a sphere of equivalent volume. The radius of this cell is related to sink concentration ρ_c by relationship

$$\frac{4}{3}\pi R^3 \rho_c = 1 \tag{3.65}$$

This assumption makes the problem spherically symmetric.

Concentration of mobile particles $\rho(r,t)$ changes periodically when passing from one cell to another. It then follows that function $\rho(r,t)$ should have a zero derivative on the boundary of the cell—a flux through outer boundary R of the equivalent sphere equals zero:

*The size of the mobile particles must be much smaller than the distance between traps.

$$\left.\frac{\partial \rho(r, t)}{\partial r}\right|_{r=R} = 0 \tag{3.66}$$

Now Equation (3.62) takes the form

$$r \frac{\partial \rho(r, t)}{\partial t} = D \frac{\partial^2(\rho r)}{\partial r^2} \tag{3.67}$$

and it must be solved with boundary conditions (3.66) and (3.63). In the latter of these conditions it can be assumed that (without loss of generality) $\rho_s = 0$, so

$$\rho(a, t) = 0 \tag{3.68}$$

The initial distribution in a cell will be considered uniform, i.e.,

$$\rho(r, 0) = \rho_0 \tag{3.69}$$

The solution of Equation (3.67) will be sought in the form of expansion in eigenfunctions of equation

$$\nabla^2 \psi + \lambda_n^2 \psi = 0$$

i.e.,

$$\rho(r, t) = \sum_{n=0}^{\infty} a_n \psi_n(r) \tag{3.70}$$

Substituting (3.70) into (3.67), one derives an equation for determining coefficients a_n:

$$\frac{da_n}{dt} = -D\lambda_n^2 a_n(t)$$

Then

$$\rho(r, t) = \sum_{n=0}^{\infty} \bar{a}_n \psi_n(r) \exp(-t/\tau_n) \tag{3.71}$$

where

$$\tau_n = (\lambda_n^2 D)^{-1}$$

Coefficients $\bar{a}_n$ of expansion (3.71) are determined by integrating over the volume of sphere V_c:

$$\bar{a}_n = \int_{V_c} \psi_n \rho(r,0)dV = 4\pi\rho_0 a C_n \lambda_n^{-2}$$

where C_n are normalizing multipliers of eigenfunctions ψ_n:

$$\psi_n = \frac{C_n}{\lambda_n} r^{-1} \sin [\lambda_n (r - a)],$$

$$\frac{C_n^2}{\lambda_n^2} = \frac{1}{2\pi} \frac{1 + \lambda_n^2 R^2}{\lambda_n^2 R^2 (R - a) - a} \tag{3.72}$$

Conditions (3.66) lead to an equation defining eigenvalues λ_n:

$$\tan [\lambda_n (R - a)] = \lambda_n R, \qquad n = 0,1,2,\ldots \tag{3.73}$$

The lowest root λ_0 can be evaluated in the form of a series in small parameter ξ aR^{-1}, assuming that $\lambda_0 R \ll 1$:

$$\lambda_0^2 = 3\xi R^{-2}[1+ \frac{9}{5} \xi + 0(\xi^2)]$$

To find the other roots, Equation (3.73) is solved graphically. The estimate can be obtained assuming $\lambda_n R \gg 1$. Then

$$\lambda_n \approx \pi n/(2R) \qquad \text{at} \qquad n \gg 1$$

(the value of the second root appeared to equal $\lambda_1 = 4.5/R$ [169]). To clarify the competing influence of sinks on the kinetics of mobile particle decrease, we obtain an expression for flux $j(t)$ into one sink—similar to the Smoluchowski Method. Thus, comparing the solutions of this problem with the results of Smoluchowski reveals the effects of trap competition.

Value $j(t)$ is easily found from (3.71) and (3.72):

$$j(t) = 4\pi Da \sum_{n=0} \bar{a}_n C_n \exp(-t/\tau_n) \tag{3.74}$$

It is clear that the principal contribution into this expression is made by several first terms. This follows from writing approximate expressions for τ_n and $\bar{a}_n C_n$:

$$\tau_0 = \lambda_0^{-2} D^{-1} = R^3 (3aD)^{-1}; \qquad \text{at} \qquad n \gtrsim 1,\ \tau_n \sim R^2 (D\pi^2 n^2)^{-1}$$

$$\bar{a}_0 C_0 = \rho_0 [1 + 0(\xi)]; \qquad \text{at} \qquad n \geqslant 1,\ \bar{a}_n C_n \sim 2\rho_0 \xi$$

Thus all the components in (3.74) corresponding to $n \geqslant 1$, are proportional to ξ, i.e., they are small in absolute value (it is assumed that $\xi \ll 1$); at this $\tau_n \ll \tau_0$ at $n = 1, 2, \ldots$ Therefore upon expiration of a time larger than $\tau_R = R^2 D^{-1}$—the time of a "diffusional passage" of outer sphere radius R (in other words at $t > \tau_R$) only the first two terms can be considered in (3.74). At very small times, when $t \ll \tau_R$, all the terms in the series must necessarily be taken into account, but in this case the problem becomes analogous to the problem of a single sink ($R \to \infty$). From that point revealing the role of sink competition is of little interest (at small times the role is insignificant).

The general conclusion on the kinetics of the process follows from comparing expressions (3.71) and (3.74): the flux is not a linear function of the concentration. Particle concentration $\rho(t)$ in a cell at time t can be easily calculated:

$$\rho(t) = 4\pi a \sum_{n=0} \lambda_n^{-2} \bar{a}_n C_n \exp(-t/\tau_n) \tag{3.75}$$

It is seen that the kinetics of the process cannot be described by an expression imposed by the classical kinetics of monomolecular reactions

$$\frac{d\rho_{cl}}{dt} = -k_1 \rho_{cl} \tag{3.76}$$

i.e., the rate constant is a time function. Now let us analyze the expression for flux $j(t)$ having confined ourselves to two terms

$$j(t) \approx 4\pi D a \left(e^{-t/\tau_0} + \frac{2a}{R} e^{-t/\tau_1} \right) \rho_0$$

Separating an interval t when $\tau_0 \gg t \sim \tau$, the flux can be presented in the form

$$j(t) = 4\pi D a \rho_0 \quad 1 + \frac{2a}{R} e^{-t/\tau_1}$$

However time $\tau_1 \gg a^2/D$. This means that unlike the result of Smoluchowski, the obtained flux is nonstationary for a long time comparable to τ_1. At $t \gg \tau_1$, leaving only the first term in (3.74) or (3.75), one obtains:

$$\rho(t) = \rho_0 \exp(-t/\tau_0), \quad j = k_D \rho(t)$$

i.e., the reaction assumes the usual monomolecular character. Thus the competition of sinks prolongs the non-stationarity of rate constant $k(t)$

up to times comparable to the times for a diffusional passage through sphere of radius R. At this $k(t) > k_D = 4\pi Da$.

As it is seen, the qualitative picture of the kinetics of Brownian particle absorption on sinks obtained in the Method of Cells (at times $t \gg \tau_1$) does not differ dramatically from the results obtained by the Smoluchowski Method, though the latter does not include an assumption on the regularity of trap distribution. Therefore it might have appeared possible to assert that the regularity of the sink lattice does not play any significant role in the three-dimensional case. However, this conclusion is premature, since at large times the kinetics can be determined by the fluctuations in the density of particles and traps. A general analysis of the problem of diffusion in the presence of traps is given in papers [28,174–176] and in Section 3.6. Now let us discuss a problem where correlation is displayed more explicitly, namely the two-dimensional variant of the above-considered problem.

In the case of cylindric symmetry, Equation (3.62) take the form

$$\frac{\partial \rho(r,t)}{\partial t} = D\left(\frac{\partial^2 \rho}{\partial r^2} + \frac{1}{r}\frac{\partial \rho}{\partial r}\right) \tag{3.77}$$

It is solved with conditions (3.66), (3.68) and (3.69)

$$\rho(r,0) = \rho_0, \qquad \rho(a,t) = 0, \qquad \partial\rho/\partial r\big|_{r=R} = 0$$

flux $j(t)$ being prescribed by formula

$$j(t) = 2\pi a \partial\rho/\partial r\big|_{r=a} \tag{3.78}$$

Applying Laplace transform

$$\rho(r,s) = \int_0^\infty \rho(r,t)e^{-st}dt$$

one obtains equation $s\rho(r,s) - \rho_0 = D(\rho'' + r^{-1}\rho')$ instead of (3.77). Its solution for satisfying the boundary conditions can be presented in the form

$$\rho(x) = \frac{\rho_0}{s}\left[1 - \frac{I_0(x)K_1(x_R) + K_0(x)I_1(x_R)}{I_0(x_0)K_1(x_R) + K_0(x_0)I_1(x_R)}\right]$$

where $x = r\sqrt{s/D}$; $x_0 = a\sqrt{s/D}$; $x_R = R\sqrt{s/D}$; $I_n(x)$ and $K_n(x)$ are Bessel functions of the imaginary argument.

Using relationships

$$\frac{d}{dx}I_0(x) = I_1(x), \qquad \frac{d}{dx}K_0(x) = -K_1(x)$$

one finds an expression for flux (3.78):

$$j(s) = -\frac{2\pi x_0}{s} \rho_0 D \frac{I_1(x_0)K_1(x_R) - K_1(x_0)I_1(x_R)}{I_0(x_0)K_1(x_R) + K_0(x_0)I_1(x_R)} \tag{3.79}$$

This expression is difficult to analyze in the general form, so let us investigate its asymptotic behavior, assuming that $x_R \ll 1$ which corresponds to times $t \gg R^2 D^{-1} \sim \tau_1$.

The asumptotic behavior of Bessel functions is determined by expressions ($x \ll 1$):

$$K_1(x) \approx x^{-1} + \frac{x}{2} \left[\ln \frac{x}{2} - \psi(1) - \psi(2) \right] + 0(x^3 \ln x)$$

$$K_0(x) \approx - \ln \frac{x}{2} + \psi(1) + 0(x^2 \ln x)$$

$$I_0(x) \approx 1 + \frac{x^2}{4} + 0(x^4); \qquad I_1(x) \approx \frac{x}{2} + 0(x^3) \tag{3.80}$$

where $\psi(z)$—is the logarithmic derivation of gamma function; $\psi(1) = -\gamma$; $\psi(2) = 1 - \gamma$; γ is the Euler constant ($\gamma = 0.5772$).

Taking into account (3.80), one can derive an approximate expression for flux $j(s)$ (3.79):

$$j(s) \approx \pi x_0 D s^{-1} \rho_0 R a^{-1} x_R \left[1 + \frac{1}{2} x_R^2 \left(\ln \frac{R}{a} - \gamma \right) \right]^{-1}$$

Now flux $j(s)$ should be brought into correlation with particle density on the outer surface of the sphere (as required by the introduction of the effective rate constant):

$$\rho(x_R) = \rho_0 s^{-1} \frac{x_R^2}{2} \left(\ln \frac{R}{a} - 1 \right) \left[1 + \frac{x_R^2}{2} \left(\ln \frac{R}{a} - \gamma \right) \right]^{-1} \tag{3.81}$$

where the relationship for the Bessel function is used

$$I_\nu(x) K_{\nu+1}(x) = \frac{1}{x} - I_{\nu+1}(x) K_\nu(x)$$

It is clear that the flux into the trap is proportional to particle density on the outer surface of a cell (3.81):

$$j(s) = 2\pi D \left[\ln \frac{R}{a} - \gamma \right]^{-1} \rho(x_R) \tag{3.82}$$

Applying the reverse Laplace transform, we shall find dependence $\rho(R, t)$:

$$\rho(R, t) = \rho_0 \exp \left\{ - \frac{2Dt}{R^2 \left[\ln \frac{R}{a} - \gamma \right]} \right\}$$

and thus the time dependence of flux $j(t)$:

$$j(t) = 2\pi D\rho_0 \left[\ln \frac{R}{a} - \gamma\right]^{-1} \exp\left\{-\frac{2Dt}{R^2\left[\ln \frac{R}{a} - \gamma\right]}\right\} \tag{3.83}$$

Applying the separation of variables similar to that in the three-dimensional problem, we should have obtained value $\lambda_0^2 = 2R^{-2}[\ln R/a - \gamma]^{-1}$ for the exponent which exactly coincides with the value obtained by Ham [169]. A rigorous solution of the problem [174] gives the same value of the first root. Thus, the Method of Cells describes well the kinetics of absorbing diffusing particles on traps in the case of a regular distribution of the latter ones.

Several conclusions follow from relationship (3.83). There is a time region $a^2D^{-1} \ll R^2D^{-1} \sim t \ll R^2D^{-1} \ln R/a$ when $j(t)$ can be practically considered a constant. The surface reaction can be described by the classical kinetics of monomolecular reactions (3.76) with value $k_1 = 2\pi D_{ef}\rho_c$. It is also clear that the Smoluchowski Method cannot be applied [50] in the considered two-dimensional case since $j(t) \to 0$ at $R \to \infty$. The competition of sinks leads to the finite value of the flux. The course of the reaction renormalizes the diffusion coefficient:

$$D_{ef} = D\left[\ln \frac{R}{a} - \gamma\right]^{-1}$$

the value of which is much smaller than the real one at (R/a) i.e., at low concentrations of sinks. Since the value of diffusion coefficient D is independent of trap concentration, and the traps themselves are immobile, the emergence of the effective diffusion coefficient can be connected with the correlation of mobile particles due to absorption on sinks.

3.4. THE EFFECTIVE MEDIUM METHOD

As it is quite difficult to consider the problem in the general case taking into account particle correlation and the competition of sinks in the case of their irregular arrangement, different simplified methods based on the idea of "an effective medium" have been developed [175–177]. Let us discuss one of these methods connected with both the Method of Intervals and the Method of Cells which is widely used in analyzing the electrical properties of disordered many-component materials [178], and in analyzing the diffusion of defects to absorbing sinks [167,179–182]. In the above-mentioned papers, the actual system which consisted of diffusing active centers and arbitrarily arranged sinks, is replaced by an effective absorbing medium with a uniform trap distribution. The characteristic parameters of the effective medium (force of sinks) are sought using a self-consistent procedure.

We shall elucidate the way to determine the characteristic parameters of the method for the case of spheric sinks of radius a. Let us consider flux J_a of diffusing particles into a trap, located in spheric region I, free of other sinks, and having radius R. The whole sphere is placed into absorbing medium II. Then it is natural to equate the found flux J_a into one trap multiplied by sink concentration c_a to the full flux of particles disappearing on the traps proportional to absorption parameter α^2. In the general case, the traps can not only absorb active particles but also emit them. The capability of traps to emit particles will be characterized by parameter K_1, which signifies the emission rate per time unit. Values α^2 and K_1 are the parameter of the absorbing effective medium.

Let us consider a stationary problem for a system composed of diffusing particles of one type, traps of different type and a source of particles. Then in region I around the absorbing sink

$$Dr^{-2}\frac{d}{dr}\left(r^2\frac{dC_I}{dr}\right) + K, \qquad a \leqslant r \leqslant R \tag{3.84}$$

under boundary condition

$$D\left.\frac{dC_I}{dr}\right|_{r=a} = \bar{K}(c_I - c_1) \tag{3.85}$$

where $c_1(r)$ is the concentration of active centers in region I; K is the rate of particle generation; $\bar{K}$ is the rate of transfer through the surface of the effective medium matrix; and c_1 is the equilibrium particle concentration.

In region II, the corresponding equation should read as

$$r^{-2}D\frac{d}{dr}\, r^2\frac{dc_{II}}{dr^2} + K' - Dd^2c_{II}(r) = 0 \tag{3.86}$$

where c_{II} is the concentration of particles in region II; K' is the effective rate of particle generation in the medium.

If there are n types of traps, each of which is characterized by parameter α_i^2 called the force of the sink, then

$$\alpha^2 = \sum_{i=1}^{n} \alpha_i^2$$

$$K' = K + K_1 + K_n$$

where K_1 is the parameter characterizing the increase in the rate of particle generation due to emission from sinks; K_n are analogous coefficients from other $(n - 1)$ types of sinks.

At boundary $r = R$ the continuity of concentration and its derivative must be required, i.e.,

$$C_I(R) = c_{II}(R) \tag{3.87}$$

$$\frac{\partial C_I}{\partial r} = \frac{\partial c_{II}}{\partial r} \quad \text{at} \quad r = R \tag{3.88}$$

The solution of system (3.84–3.86) has been discussed in detail in papers [179,180]. In the simplest case, when sink concentration is low, i.e.,

$$\frac{a}{R} \ll 1, \qquad \alpha R \ll 1 \tag{3.89}$$

the solution of the equations under conditions (3.87) and (3.88) has the form

$$c_I = \mu - \lambda a r^{-1} - K r^2 D^{-1}$$

$$c_{II} = K' D^{-1} \alpha^{-2} + A R r^{-1} \exp\left[-\alpha(r - R)\right] \tag{3.90}$$

where parameters μ, λ, A under condition (3.89) can be presented in the form

$$\mu = KR^2(2D)^{-1} + K'\alpha^{-2}D^{-1}, \qquad \lambda = \mu - c_1, \qquad A = KR^3(3D)^{-1} - a\lambda R^{-1}$$

The rate of particle decrease on a sink in region II will be determined using (3.90):

$$J_a = 4\pi D a^2 \left.\frac{dc_I}{dr}\right|_{r=a} = 4\pi D \lambda$$

The condition of self-consistency is now the following

$$J_a c_a = -K_1 + D c_{II}(\infty) \alpha_1^2 \tag{3.91}$$

where $c_{II}(\infty) = K'/D\alpha^2$ is the mean stationary concentration of particles.

Using the expression for J_a on the basis of relation (3.91) one obtains

$$K\alpha_1^2 - K_1\alpha_n^2 + K_n\alpha_1^2 = 4\pi a c_a [K + K_1 + K_n - D\alpha^2 c_1]$$

The comparison of the coefficients in the left and right parts of this equation gives two relations

$$\alpha_1^2 = 4\pi a c_a$$

$$-K_1\alpha_n^2 + K_n\alpha_1^2 = 4\pi a c_a (K_1 + K_n - D\alpha^2 c_1)$$

from which it follows that

$$K_1 = 4\pi D c_a c_1$$

Thus, it is possible to define necessary parameters α_1^2 and K_1 of the effective medium with an accuracy up to $0(a/R)$. In principle, these parameters can be obtained also in the general case without requirement (3.89). It is important for us to stress that the comparison of the exact and approximate solutions when $R = a$ shows a good coincidence. But as can be easily seen from (3.84) and (3.86) in the case $R = a$, the Method of the Effective Medium is equivalent to the Method of Intervals.

In conclusion, it should be noted that the methods considered (of intervals and of the effective medium) permit taking into account correlation effects which are important for analyzing the kinetics of various physical-chemical processes. However, the applicability of these methods is restricted by the assumption on the low density of reagents when sufficient information comes from two-particle distribution functions. To overcome this restriction, more rigorous methods should be applied which use correlation functions of second and higher orders. Let us consider one of these methods which, in principle, allows investigation of DCR kinetics in the case of arbitrary relationships of the components' concentrations.

3.5. THE METHOD OF GREEN'S FUNCTIONS [183]

3.5.1. General Considerations

Let us consider the method of a theoretical description of DCR kinetics which uses averaging over reciprocal configurations of particles and takes into account the redistribution of the diffusional flux between neighboring particles. We shall illustrate the method by analyzing the kinetics of monomolecular reaction

$$A + B \xrightarrow{k} C$$

Particles of both type are assumed to be spherical and for convenience B type particles—immobile.

We shall seek for the average (over volume V of the reaction system) value of function $\rho(r, t)$, which is the solution of diffusion equation

$$\partial \rho(r, t)/\partial t = D\Delta \rho \tag{3.92}$$

with boundary conditions on surface Σ, which is the sum of the reaction surfaces of B particles. The surface of i-particle is given by condition $|r - R_i| = R$ (where R is the radius of the reaction surface; R_i is the radius-vector of the center of an i-particle B; $i = 1,2,\ldots,N$; N is the general number of B particles in the system).

Let us assume that A type particles are created instantly at initial time moment $t = 0$ with density ρ_0 constant throughout the volume of the system. Let the reaction be controlled by diffusion, i.e., $k \gg k_D$. In this case the boundary and initial conditions should be written in the following way

$$\rho(r,0) = \rho_0, \qquad \rho[r(\Sigma_i),t] = 0 \tag{3.93}$$

Application of the Laplace transform to Equations (3.92) and (3.93) gives

$$(x^2 - \Delta)p(r,s) = \rho_0 D^{-1} \tag{3.94}$$

$$\rho[r(\Sigma_i),s] = 0 \tag{3.95}$$

where

$$\rho(r,s) = \int_0^\infty e^{-st}\rho(r,t)\,dt, \qquad x^2 = s/D$$

Using direct substitution it can be proved that function

$$\rho(r,s) = \frac{\rho_0}{s} + \sum_{i=1}^{N} c^i(x)G(x\,|r - R|) \tag{3.96}$$

is the solution of Equation (3.94) at any values of $x(0 < x < \infty)$ and at all r, except the ones coinciding with the center of any B particle: $r \neq R_i$, $i = 1,2,\ldots,N$. Here $G(x,r)$ is Green's function of operator $(x^2 - \Delta)$: $(x^2 - \Delta)G(x,r) = \delta(r)$, $C^i(x)$ is an arbitrary function of x, corresponding to i- B particle. Boundary condition (3.95) permitting, in principle, writing N equations for determining N unknown functions $C^i(x)$. Each of them, because of density redistribution (of diffusional flux) between neighboring B particles, depends on x and in addition on $(N - 1)$ spatial (vector) variables—positions of B particles relatively to the i-th one:

$$c^i(x) = C(x, R_1 - R_i, \ldots, R_N - R_i)$$

Mean density $\rho(r_0,s)$ in arbitrary point r_0 of an actual system can be obtained by averaging Equation (3.96) over all the possible configurations of B particles. In particular, in the case of random distribution of the absorbing centers, one can derive

$$\rho(r_0,s) = \int \rho(r_0,s,R_1,\ldots,R_i,\ldots R_N)\prod_{j=1}^{N}\frac{dR}{V} = \frac{\rho_0}{s} +$$

$$+\frac{1}{V}\sum_{i}^{N}\int dR_i C_1^{\,i}(x)G(x\,|r_0 - R_i|) = \frac{\rho_0}{s} + \frac{N}{V}\int C_1^{\,1}G_{01}dR_1$$

where

$$G_{01} = G(x|R_1 - r_0|);$$

$$C^i(x) = \int \prod_{j \neq i} \frac{dR_j}{V} C(x, \ldots, R_j - R_i, \ldots) \equiv C_1^{\ 1}$$

Function G_{01} takes into account the influence of a B particle located in point R_1 on the density of mobile particles in point r_0. Function C_1^1 is obtained upon averaging function $C^i(x)$ and represents a correlation function of first order.

Let us denote the correlation function C_K^i, $i \leqslant k$ of k-order as a result of averaging the weighting function

$$C^i = C(x, r_1 - r_i, \ldots, r_K - r_i, R_{K+1} - r_i, \ldots, R_N - r_i)$$

(where $R_j \equiv r_j$ at $1 \leqslant j \leqslant k$) over coordinates R_j, $k < j \leqslant N$ of all B particles besides the k first particles located in points $r_1, r_2, \ldots, r_K$.

Writing the boundary condition (equaling the concentration to zero) on each of k particle B and averaging the derived equation over the coordinates of the rest of the particles, gives the following system of equations:

$$\frac{\rho_0}{s} + \sum_{j=1}^{K} C_K^{\ j} G_{ij} + \frac{N - K}{V} \int C_{K+1}^{\ K+1} G_{i,K+1} dR_{K+1} = 0$$

$$i = 1, 2, \ldots, K \tag{3.97}$$

Function G_{ij} takes into account the effect of the j-th B particle upon density $\rho(r,s)$ on the surface of the i-th B particle ($i \neq j$) while function G_{ii}, the influence of the i-th particle upon the density on its eigen reaction surface

$$G_{ij} = G(x|r_j - r_i|), \qquad G_{ii} = G(xR)$$

The system of Equation (3.97) permits obtaining a closed equation with respect to all functions C_K^i, $1 \leqslant i \leqslant k$, $1 \leqslant k \leqslant N$, and hence determines average density $\rho(r_0, s)$, i.e., to obtain the formally exact solution of the problem.

In the majority of the systems, general number N of B particles and volume V of the system can be considered infinitely large, so

$$\lim_{V \to \infty} \frac{N - K}{V} = n \qquad \text{at any finite } k$$

Value n is the parameter of the system; using (3.96) and (3.97) one can obtain the following expansion of density $\rho(r_0, s)$ as a series in this parameter:

$$\rho(r_0,s) = \frac{\rho_0}{s}\left[1 - n\int \frac{G_{01}}{G_{11}}\,dr + n^2 \iint \frac{G_{01}\begin{vmatrix} G_{11} & 1 \\ G_{21} & 1 \end{vmatrix}}{\begin{vmatrix} G_{11} & G_{12} \\ G_{21} & G_{22} \end{vmatrix}}\,dr_1 dr_2 - \right.$$

$$- n^3 \iiint \frac{G_{01}\begin{vmatrix} G_{11} & G_{13} \\ G_{12} & G_{23} \end{vmatrix}\begin{vmatrix} G_{11} & G_{12} & 1 \\ G_{21} & G_{22} & 1 \\ G_{31} & G_{32} & 1 \end{vmatrix}}{\begin{vmatrix} G_{11} & G_{12} \\ G_{21} & G_{22} \end{vmatrix}\begin{vmatrix} G_{11} & G_{12} & G_{13} \\ G_{22} & G_{22} & G_{23} \\ G_{31} & G_{32} & G_{33} \end{vmatrix}} \times$$

$$\left. \times\, dr_1 dr_2 dr_3 + \ldots \right] = \frac{\rho_0}{s}\left[1 + \sum_K (-n)^K \int \prod_{i=1}^{K} dr_i \times \right.$$

$$\left. \times \frac{\prod_{i=1}^{K} P_i}{\prod_{i=1}^{K} D_i} \right] \qquad (3.98)$$

The coefficient at k-term of expansion (3.98) is a k-fold integral over the positions of the first k particles. The corresponding integrand expression is equal to the quotient obtained when the product of k determinants of type P_i is divided by the product of k determinants of type D_i, $1 \leqslant i \leqslant K$. Determinant D_i is a determinant of i-order, where at the intersection of the l-line and j-column stands function G_{lj}. Determinant P_i, $i < k$ is obtained from determinant D_i by substituting the last i-column for a column $\{a_l\}$ containing functions $a_l = G_{l,\,i+1}$ (instead of function G_{li}, in the corresponding column of determinant D_i). At $i = k$ each element of the last k-column of determinant P_k equals 1. Diagonal terms of determinants D_i equal functions $G_{ll} = G(xR)$. In determinants P_i the diagonal terms in all columns besides the last one also equal $G(xR)$. The expression for $\rho(r_0,s)$ is an exact Laplace-transform of average reagent density. Let us consider it in more detail in the example of three-dimensional systems when influence function $G(y)$ is determined by the expression $G(y) = 1/y \exp(-y)$.

Substituting the value of $G(y)$ into the expression for $\rho(r_0,s)$, one transforms the latter into a dimensionless form. Now let us multiply all columns of each of D_i determinants into R. In determinants P_i each column except the last one is also multiplied by R, while the last one is divided by x. For the coefficient at the k-term of series (3.98) to remain unchanged, it is necessary to multiply each term of the series by $(xR)^K$. Then the dimensionless variable of the integration is equal to:

$$y_i = xr_i, \qquad dy_i = x^3 dr_i$$

The final expansion of function $\rho(r_0,s)$ as a series in the density of B particles can be written in the following way:

$$\rho(r_0,s) = \frac{\rho_0}{s}\left[1 + \sum_{K=1}^{\infty} (-\chi)^K f_K(a)\right] \tag{3.99}$$

where

$$\chi = 4\pi n D R s^{-1}, \qquad a = xR = R\sqrt{s/D} \tag{3.100}$$

$$f_K(a) = (4\pi)^{-(K+1)} \int\cdots\int \prod_{i=1}^{K} P_i \left(\prod_{i=1}^{K} D_i\right)^{-1} dy_1 \ldots dy_{K+1} \tag{3.101}$$

$$D_i = \begin{vmatrix} e^{-a} & aG_{12} & aG_{13} \ldots aG_{1i} \\ aG_{21} & e^{-a} & aG_{23} \ldots aG_{2i} \\ \cdots & \cdots & \cdots \\ aG_{i1} & aG_{i2} & aG_{i3} \ldots e^{-a} \end{vmatrix} \tag{3.102}$$

$$P_i = \begin{vmatrix} e^{-a} & aG_{12} \ldots\ldots aG_{1i-1} & G_{1i+1} \\ aG_{21} & e^{-a} \ldots\ldots aG_{2i-1} & G_{2i+1} \\ \ldots & \ldots & \ldots \\ aG_{i1} & aG_{i2} \ldots\ldots aG_{ii-1} & G_{ii+1} \end{vmatrix} \qquad (3.103)$$

Let us clarify the last relation for determinant P_i. As it follows from Equation (3.98), each element of the last column is equal to I in determinant P_k, and the integrand itself is a function of k variables $y_1, y_2, \ldots, y_k$. Accordingly, integration in (3.98) should be carried out with respect to k variables. However, since

$$(4\pi)^{-1} \int G(y)dy = 1$$

expression (3.101) will not change if for generality of the writing the unity in the last column of determinant P_k is substituted for function $G_{l,\,K+1}$ (l is the number of the line) and if the obtained expression for f_K is integrated with respect to y_{K+1} as has been done in writing expression (3.101).

In the quasi-stationary regime $t \gg R^2/D$ (or at $s \ll D(R^2)$, parameter a may be considered small which provides for the following expansion of function $\rho(r_0,s)$ determined by the corresponding expansion of coefficients f_K:

$$\rho(r_0,s) = \frac{\rho_0}{s}\left[1 + \sum_{K=1}^{\infty} (-\chi)^K f_K(a)\Big|_{a=0}\right] + $$

$$+ \frac{\rho_0 a}{s} \sum_{K=1}^{\infty} (-\chi)^K \frac{df_K(a)}{da}\Big|_{a=0} + \ldots \qquad (3.104)$$

Each element comprising the matrix of determinant D_i, besides the diagonal element, contains small parameter a as a multiplier [see (3.104)]. Similarily, each matrix element of determinant P_i (besides the diagonal elements of the last column) also contains parameter a. Consequently, the following expressions are valid, accurate up to values of second and greater order in a:

$$D_i = 1 - ia + \ldots$$

$$P_i = G_{ii+1} - a\left[(i-1)G_{ii+1} + \sum_{l=1}^{i-1} G_{il}G_{li+1}\right] + \ldots$$

Substituting determinants D_i and P_i in f_K for their zero order approximation gives

$$f_K(a)\Big|_{a=0} = (4\pi)^{-(K+1)} \int\!\ldots\!\int \prod_{i=1}^{K} G_{ik}dy_1\ldots dy_K \tag{3.105}$$

Proceeding to new variables $\eta_{i+1} = y_{i+1} - y_i$, $i = 1,2,\ldots,K$ in integrating (3.105) one obtains

$$f_K(a)\Big|_{a=0} = (4\pi)^{-(K+1)} \int\!\ldots\!\int \prod_{i=1}^{K} G(|\eta_i|)d\eta_1\ldots d\eta_{K+1} = 1 \tag{3.106}$$

Substituting expression (3.106) into equation (3.104), one deduces the value of function $\rho(r_0,s)$ in zero order approximation with respect to parameter a:

$$\rho(r_0,s)\Big|_{a=0} = \rho_0/s(\chi+1)$$

It can be shown [183] that in the general case function $\rho(r_0,s)$ may be presented in the form

$$\rho(r_0,s) = \rho_0/s\{1 + \chi[1 + g(a,\chi)]\}$$

where function $g(a,\chi)$ is determined only by the form of function $G(y)$. Therefore, the expansion of function $\rho(r_0,s)$ in a parameter a (and thus also in parameter $d_n = 4\pi R^3 n$) is determined by the corresponding expansion of function $g(a,\chi)$:

$$g(a,\chi) = g(a,\chi)\Big|_{a=0} + a\,\frac{\partial g(a,\chi)}{\partial a}\Big|_{a=0} + \ldots \tag{3.107}$$

and each term of expansion (3.107)

$$g_m \equiv \frac{\partial^m g(a,\chi)}{\partial a}\Big|_{a=0}, \qquad m > 1$$

is a small value of order α_n relative to the foregoing term g_{m-2}.

Thus, to a small accuracy of parameter α_n terms, the mean density is equal to

$$\rho(r_0,s) = \rho_0/s[1 + f(1 + a\sqrt{f+1})]$$

Applying the inverse Laplace transform one obtains:

$$\rho(r_0,t) = \rho_0 \exp[-\tau_n(1-\alpha_n)] \operatorname{erfc}(\sqrt{\alpha_n \tau_n})$$

where

$$\tau_n = 4\pi R D n t$$

The first term of the expansion of function $\rho(r_0,t)$ in variable $\alpha_n\tau_n$ coincides with the first term of the expansion of function

$$\rho(t) = \rho_0 \exp\left[-n \int_0^t k_{ex}(\tau)d\tau\right]$$

where

$$k_{ex}(t) = 4\pi D R(1 + R/\sqrt{\pi D t})$$

This means that at short times $\tau_n \ll \alpha_n^{-1}$, when B particles may be considered isolated, the obtained time dependence of the mean concentration coincides with the known solutions (see Section 3.2). However, at large times when the competition between various B particles becomes significant, the obtained solution may differ considerably from the results of problems on the basis of two-particle distribution functions.

3.5.2. Application of Green's Method to Describing Heterogeneous Diffusion Controlled Reactions [184]

When investigating the influence of the diffusion rate of reagents upon the kinetics of heterogeneous processes of chemical reactions, the reaction surface is usually considered to be an infinite uniform plane [3,185,186]. Such an assumption holds if the atoms comprising the matrix of the solid phase are one of the reagents (reagent B). In the opposite case, when reagent B is impurity particles of characteristic size R randomly distributed with low density $n(nR^2 \ll 1)$ on the surface of the solid, such an assumption generally speaking cannot be applied.

Let us analyze the latter situation. Evidently in this case the process can be conditionally divided into two stages. In the initial stage, the reaction can be entered only by those particles of the dissolved-in-liquid phase reagent A which were initially located quite close to the reaction surface; and active B particles may be considered isolated [187]. At large times $t > T \approx R_m^2/D$ (where R_m is the mean distance between B particles) the reaction is entered by A particles initially separated from the surface by a distance much larger than R_m. The kinetics are determined by the diffusion of reagent A to the solid phase surface (it not being possible to consider absorbing centers to be isolated due to the correlation in the particle positions: each A particle can interact with any B Particle).

The problem can be made more specific in the following way [184]. Let there be randomly distributed centers of spheric particles of reagent B on infinite plane XY (impenetrable for component A). A reversible bimolecular reaction proceeds on the surface of each particle. Initially reagent A is uniformly distributed with density ρ_0 in half-space $z \geqslant 0$. To find the density of reagent A in such a system, it is necessary to solve Equation (3.92) with initial condition (3.93) and boundary condition

$$x_1 \frac{\partial \rho}{\partial r}\bigg|_{r \text{ belomgs to } \Sigma_i} = k_1 \rho(R_{\Sigma_i}, t) - k_{-1} \tag{3.108}$$

where $i = 1, 2, \ldots N$; N is the total number of particles; k_1 and k_{-1} are the rate constants of the direct and reverse reactions, respectively.

The flux of component A through parts of surface XY, devoid of B particles, is equal to zero. The reaction surface of each B particle is a half sphere of radius R, so $x_1 = 2\pi n R^2$ in the quasi-stationary regime, weighting functions of sources $c^i(x)$ can be considered equivalent. In the case of uniform distribution of B particles on surface XY it is possible to write that

$$\rho(r_0, s) = \frac{\rho_0}{s} + n \int C(x) G(xr) dS$$

where $G(xr)$ is the influence function of the three-dimensional problem. Integration is carried out over the whole of plane XY.

The corresponding flux Φ through the surface of any B particle is, on average, equal to

$$\Phi = 2\pi D R^2 C(x) \frac{\partial G(xr)}{\partial r}\bigg|_{r=R} = 2\pi D C e^{-xh}(1 + xR)$$

At sufficiently large times $t \gg R^2/D(Rx \gg 1)$ the boundary condition can be rewritten in the following form:

$$-4\pi D C(x) = k_1 \left[\frac{\rho_0}{s} + C(x)\frac{2\pi n\sqrt{D}}{\sqrt{s}} + \frac{C(x)}{s}\right] - \frac{k-1}{s} \tag{3.109}$$

Determining function $C(x)$ from (3.109) one obtains the following expression for the Laplace transform of the reagent A flux (per surface unit of plane XY):

$$f(s) = \frac{\Phi(s)}{n} = \left(\rho_0 - \frac{k-1}{k_1}\right) \frac{k_H n \rho_0}{s + k_H n\sqrt{s/D}} \tag{3.110}$$

where

$$k_H \approx 2\pi D R k_1 / (4\pi D R + k_1)$$

Applying the inverse Laplace transform finally gives

$$f(t) = \left(\rho_0 - \frac{k_{-1}}{k_1}\right) k_H n \exp\left[k_H^2 n^2 t / D\right] \operatorname{erfc}\left(\frac{k_H n \sqrt{t}}{\sqrt{D}}\right)$$

The obtained solution coincides with the expression for the diffusion flux referred to a surface unit of the infinite plane with a boundary condition on one side of this plane

$$\left.\frac{\partial \rho(z,t)}{\partial z}\right|_{z=0} = h n \rho(0,t)$$

where $h = k_H / D$.

At short times $(R^2/D) \ll t \ll [k_H^2 n^2/D]^{-1}$ the flux referred to one particle B, with an accuracy to multiplier 1/2, coincides with the corresponding flux for an isolated particle (and consequently with the known solution for the three-dimensional case [187]):

$$\Phi = k_H \left(\rho_0 - \frac{k_{-1}}{k_1}\right) \left[1 - \frac{k_H^2 n^2 t}{D} - h \frac{k_H n \sqrt{t}}{\sqrt{D}} + \ldots\right]$$

At large times $t \gg D/(k_H n)^2$ the flux into surface unit depends only on the diffusion coefficient:

$$f(t) = (\rho_0 - k_{-1}/k_1)\sqrt{D/(\pi t)}$$

It follows from Equation (3.110) that in condensed phases "absorption capacity" h of a plane, where active centers are randomly distributed, is equal to

$$h = 2\pi R k_1 / (4\pi D R + k_1)$$

The maximum possible value $h_{max} = 2\pi R$ is the analogue of the diffusion limit for the rate constants of bimolecular reactions in three-dimensional systems $k_D = 4\pi D R$.

The model demonstrates the arising of specific correlation effects which determine the change in the dimensionality of the problem with increasing depth of the proceeding reaction. However, it should be stressed that the considered probelm itself can be of interest. Similar problems arise, for example, in theories of membrane separation of liquid and gas mixtures, when the membrane is characterized by a system of submicroscopic "penetrating holes" or pores distributed on a plane impenetrable for a mobile component. It is usually presumed that the transfer process can be quite well described by studying the distribution of the "reagent" in the vicinity of one pore and then summing up the contributions of separate pores for finding the total flux through the membrane. A similar assumption is made above to simplify analysis of the non-stationary process. However, if the penetration rate through the membrane is limited by the transport of a component towards a pore (or from a pore), and not by diffusion inside a membrane, the assumption on sink additivity is of little use. The perturbation of reagent distribution by the presence of a pore can be extended to very large distances. A considerable overlapping between diffusion zones of different pores should be observed even in case of a strong dilution, when $nR^2 \ll 1$. This type of long-range interaction arises during the process itself and is determined by both sink competition and effective change of dimensionality. It is shown in paper [188] that under certain conditions this reaction induces a deviation from additivity even at small pore concentrations. Membrane penetrability then cannot be practically expressed in the form of an expansion of the Taylor series in small parameter nR^2.

3.6. CONSIDERATION OF FLUCTUATIONS OF REAGENT DENSITIES [13,155]

3.6.1. Role of Density Fluctuations in the Kinetics of Bimolecular Reactions

Usually diffusion equations are written in the macroscopic formulation ignoring density fluctuations. In the absence of a chemical reaction, diffusion fluxes lead to complete space equalization of concentrations. However, even in the case of thermodynamic equilibrium, particle distribution is characterized by a certain level of nonuniformity

$$\overline{(N - \overline{N})^2} = \overline{N}$$

(the upper line denotes averaging over space).

These fluctuations do not play any significant role in the description of diffusion mass transfer. The situation changes dramatically if one analyzes a diffusion process with chemically interacting particles, for example, with the ones that quit in the course of the reaction. If the rate of such a chemical transformation is high while diffusion coefficients of the reagents are small (quantitative estimates will be given below), then in different parts of the considered volume the increase in transformation degree will be accompanied by an increase in the relative difference of local concentration (with respect to current concentrations) of separate components. This is a consequence of fluctuations in reagent distribution of the system at initial time $t = 0$. Evidently, such a concentration separation in the system can strongly affect the kinetics of the following process and determine the law of transformations at large times. Therefore, if chemical transformations occur in the system of diffusing particles, the right part of a diffusion equation must include a fluctuating term, which is usually not taken into account. The explicit form of this component is rather complex. The theory of random and generalized functions must be employed to define it.

We shall not write the most general equations but confine ourselves to elementary arguments for illustrating asymptotic laws that may be caused by fluctuations (thermodynamic ones included) in a system of chemically interacting particles.

This will be illustrated using two examples.

1. Reaction $A + B \to C$ between two substances A and B taken in equal concentration. If $A(r)$ and $B(r)$ are respectively the densities of A and B particles in space, then because of fluctuations (both thermodynamic and non-thermodynamic connected with the pre-history of the system) values $A(r)$ and $B(r)$ can differ. The mean-over-space values of these concentrations ($\bar{A}$ and $\bar{B}$ respectively) are equal according to the condition: $\bar{A} = \bar{B}$.

2. Monomolecular reaction $A + B \to C$ where B are traps of infinite capacity distributed randomly in space. We assume that $A(t) \ll B$ (the final stages of the process, i.e., the region of large times).

It appears that in the first case law $\bar{A} = \bar{B} \sim t^{-1}$, given by the formal kinetics of bimolecular reactions, changes at large times $t > \tau_1$ into $\bar{A} = \bar{B} \sim t^{-3/4}$. In the second case, the formal kinetic dependence for reactions of first order $\bar{A} = A_0 \exp(-kB_0 t)$ changes at $t > \tau_2$ into $\bar{A} \sim A_0 \times \exp(-\text{const}\ t^{3/5})$. The meaning of characteristic times τ_1, τ_2 will be clarified below.

The derivation of new asymptotically peculiar laws is of special interest to chemical kinetics. Classical equations which ignore fluctuations have, as it will be shown, a limited sphere of application.

1. Case $\bar{A} = \bar{B}$. Let us imagine a bimolecular composition reaction $\bar{A} + \bar{B} = C$. The initial state of the system with a high initial stationary concentration of A and B particles, which will be designated $A_0 = B_0$, can be prepared for example by thermal dissociation or photochemical decomposition of C particles.

Let us consider the disappearance of A and B after time moment $t = 0$ when the whole substance has been cooled or radiation treatment has ceased so that further C dissociation can be neglected. In the approximation of formal kinetics, we get

$$d\bar{A}/dt = d\bar{B}/dt = -k\overline{AB} \tag{3.111}$$

Then under initial condition $A_0 = B_0$ the following solution is derived

$$\begin{aligned} &\bar{A} = \bar{B} = A_0(1 + kA_0 t)^{-1} \quad \text{at} \quad t > 0 \\ &\bar{A} = \bar{B} = (kt)^{-1} \quad \text{at} \quad t \gg (kA_0)^{-1} \end{aligned} \tag{3.112}$$

If there is considerable non-uniformity in the distribution of A and B particles at $t = 0$ (for example, when the particles are arranged in pairs or bunches as in radiation-chemical experiments) appreciable deviations from equations of formal kinetics can be observed at initial stages.

Thus, if the particles are arranged in bunches the kinetics of the transformation $A + B \to C$ will at the first time moments be determined by particle density in the bunch and not by the average particle density in the whole volume. This question is treated in detail in [148,154].

How will (3.112) change if fluctuations are taken into account? The initial state is characterized by the equalities

$$\overline{(A - A_0)^2} = \overline{(B - B_0)^2} = \overline{A}_0 = \overline{B}_0$$

Since initial concentrations of A and B are either uncorrelated or possess negative correlation (initial excess of A suppressing dissociation creates a deficiency of B) the following relationship is valid for difference $(A - B)$

$$\overline{A - B} = C, \quad \sqrt{\overline{(A - B)^2}} \geqslant 2A_0$$

Let us consider the limit case of an infinitely fast reaction of A and B addition after cooling when $k \to \infty$ at $t > 0$.

It is well known that in this case A and B cannot coexist. Therefore in the region where $Z = A - B > 0$ we have $B = 0$, $A = Z$ and in the region where, on the contrary, $A = 0$, $B = -Z$. The size of regions with $Z > 0$, $Z < 0$ is obviously defined by a diffusion condition $l = \sqrt{Dt}$ since the equation for $(A - B)$ is independent of the course of the chemical reaction.

The fluctuation spectrum of value Z is such that the mean square of this value in each region equals the initial number of particles:

$$\overline{(ZV)^2} \approx A_0 V = B_0 V$$

Then, taking into account the size of region $V \sim l^3 \sim D^{3/2} t^{3/2}$ one can obtain

$$\overline{|Z|} = \sqrt{A_0 V^{-1}} = A_0^{1/2} D^{-3/4} t^{-3/4} \tag{3.113}$$

In the case of an instantaneous reaction in certain regions $A = |Z|$, in othe $B = |Z|$; and consequently

$$\overline{A} = \overline{B} = \frac{1}{2} \overline{|Z|} \tag{3.114}$$

Formula (3.113) is applicable only at $\sqrt{Dt} > A_0^{-1/3}$, when the diffusion path exceeds the mean distance between particles A and B in the initial state. Thus in the region of applicability of (3.113) $A \ll A_0$ as it should be.

We shall especially note the slow decrease of concentration according to law $t^{-3/4}$. If the reaction is not instantaneous, i.e., constant k is finite, then at the beginning of the process the concentration calculated using classical Equation (3.112) appears to be higher than the one calculated from (3.113), (3.114). This means that at the beginning the rate of the chemical reaction is limited. However, in the course of time a slower decrease law (3.113) will necessarily lead to a reaction regime limited by the diffusion of fluctuation. The transition from (3.112) to (3.113) and (3.114) occurs at $\tau_1 = D^3/A_0{}^3k^4$. The formal derivation of an expression for $\overline{A} = 1/2\ |\overline{Z}|$ is given in the following way.

The kinetics of equalization of fluctuation $Z(r, t)$ is described by diffusion equation

$$\frac{\partial Z}{\partial t} = D\Delta Z(r, t) \tag{3.115}$$

with initial condition

$$Z(r, 0) = Z_0(r)$$

where $Z_0(r)$ is a random quantity with Gaussian distribution $W = N \exp \times \times [-(1/2\ A_0) \int Z_0{}^2(r)dV]$, so that the mean square value of fluctuations satisfies the usual thermodynamic relation [189]; N is the normalization constant.

With use of integral representation

$$|u| = \frac{2}{\pi} \int_0^\infty \xi^{-2}(1 - \cos \xi u)d\xi$$

function $|Z(r, t)|$ can be presented in the form

$$|Z(r, t)| = \frac{2}{\pi} \int_0^\infty d\xi\xi^{-2}(1 - \cos \xi Z(r, t)) \tag{3.116}$$

The solution of Equation (3.115) can be written via Green's function for a diffusion equation

$$Z(r,t) = \int G(r, r'|t)Z_0(r')dV'$$

And then we get

$$\overline{\exp[i\xi Z(r, t)]} = \int \exp\left[i\xi \int G(r, r'|t)Z_0(r')dV'\right] W\delta Z_0 \tag{3.117}$$

Functional integral (3.117) is a Gaussian integral and can be calculated analytically. We derive

$$\overline{\exp[i\xi Z(r,t)]} = \exp\left[(-\xi^2 A_0/2)\int G^2(r,r'|t)dV'\right] \tag{3.118}$$

Substituting (3.118) into (3.116) and calculating the integral over ξ gives

$$A(t) = A_0^{1/2}/4\sqrt{\pi}(2\pi Dt)^{3/4}$$

Experimental detection of the predicted change of concentration asymptotics is quite difficult. First of all, laws t^{-1} and $t^{-3/4}$ differ only to a small extent. And besides, an experimental check of the asymptotic dependence necessarily requires providing for a disordered distribution of regions containing one component A or B to exclude long-wave components in the spectrum.

The establishment of a purely thermodynamic spectrum of fluctuations at a high temperature or upon exposure to radiation also requires time delay $t_1 > \tau_1$. Law $t^{-3/4}$ plays the role of intermediate asymptotics [189]: it works at $\tau_1 < t < t_2$ where t_2 equals either time delay t_1 or L^2/D (L is the size of the vessel which defines the longest waves in the spectrum of Z).

Finally, it should be noted that the above assumptions cannot be applied to the recombination of positive and negative ions (or ions and electrons), since electrostatic attraction truncates the spectrum of spatial fluctuations at the Debye radius of the plasma. In the presence of an additional electrolyte this limitation is removed. Thus, for example in a KNO_3 solution completely dissociated on K^+ and NO_3^-, the concentrations of ions Ag^+ and Cl^- taken in small quantities are not correlated electrostatically. Effects of the above-described type may, in principle, play an important rôle in the formation of AgCl.

2. Case $\bar{A} \ll \bar{B}$. In this case equations of formal kinetics $dA/dt = -k\bar{A}B_0$ with the Smoluchowski constant $k = 4\pi DR$ (see (1.22) yield an exponential time dependence

$$\bar{A} = A_0 e^{-ktB_0} \tag{3.119}$$

Fluctuations of the difference in the number of A and B particles are no longer important, but the type of asymptotic dependence in this case is determined by spatial density fluctuations of the number of B particles. The reason is that the system of B particles with some probability always has empty regions (voids) of large volume. Having entered such a void, particle A cannot react with B because of diffusion difficulties. The fact that A particles "get stuck" determines the asymptotic dependence of the concentration of these particles with time.

The probability that there is a void of volume V in the system of B particles is given by the Perrin formula

$$P(V)dV \approx B_0 dV \exp(-VB_0)$$

The rate of A particle decrease in such a void is determined by exponent e^{-xt}, where x is the lowest eigenvalue for the diffusion problem of an A particle in a cavity of volume V with absorbing walls:

$$xA(r) + D\Delta A(r) = 0,$$
$$A(r)\big|_{r \text{ belongs to } S} = 0 \qquad (3.120)$$

Formula (3.120) is the boundary condition of absorption on inner surface S of the isolated cavity. Since the minimal value of x at a given volume V corresponds to the spheric cavity, only spheric cavities will be taken into consideration. For such cavities $x = (\pi^2/l^2)D$ (l is the radius of the cavity), so $A(t) \approx A_0 \exp(-xt)$. Averaging over all possible l gives the following expression for density $A(t)$

$$\overline{A(t)} = 4\pi A_0 B_0 \int_0^\infty l^2 dl \exp\left(-\frac{4}{3}\pi l^3 B_0 - \frac{\pi^2 Dt}{l^2}\right) \qquad (3.121)$$

Evaluating the integral in (3.121) by the saddle-point method, one obtains:

$$\overline{A(t)} = 2^{1/5} 5^{-1/2} \pi A_0 B_0^{1/5} (\pi Dt)^{3/10} \exp\left[-\frac{5}{3}\pi^{8/5}(2B_0)^{2/5}(Dt)^{3/5}\right] \qquad (3.122)$$

It should be mentioned that the form of the pre-exponential multiplier in (3.122) depends on the choice of the form of the cavity (but of the same volume).

Thus, we obtain time dependence $\bar{A}(t) \sim \exp(-\beta t^{3/5})$ slower than the one given by equations of formal kinetics $\bar{A}(t) \sim \exp(-\gamma t)$. It is characteristic that the transfer to kinetics (3.122) is achieved at times $\tau_2 \sim (DB_0^{3/2} R^{5/2})^{-1}$ when exponents (3.122) and (3.119) become equal. The relative number of unreacted particles then depends on value $\exp \times \times [-(R^3 B_0)^{-1/2}]$ and may be not very small.

It should be noted, in conclusion, that in the case of arbitrary dimensionality d of the problem, dependence $\bar{A}(t)$ takes the form

$$\bar{A}(t) \sim \exp[-qt^{d/(d+2)}]$$

3.6.2. Asymptotics of Approaching to Equilibrium and Concentration Fluctuations in Reversible Reactions [13]

Let us consider the process of a reversible dissociation $A \rightleftarrows B + C$. The equilibrium state is determined by the law of mass action. The equilibrium constant is expressed through chemical potentials of reacting substances A, B, C. It is assumed at that each of the reagents is in a state of equilibrium at the given concentration. Consequently, it is presumed that substances A, B, C are distributed in space in accordance with the probability law of Poisson, i.e., in a way that in each small volume V, containing on the average $\bar{N} = Vn$ molecules, the root-mean-square deviation is $\Delta N = \sqrt{\overline{(N - \bar{N})^2}} = \sqrt{\bar{N}}$ fluctuations δA, δB and δc which are uncorrelated. Only in this state Gibbs energy is minimal and the mass action law $\overline{BC} = K_{eq}\bar{A}$ holds for a large volume averaged over concentrations with "thermodynamic" value K_{eq}.

As will be shown below, the kinetics of establishing full equilibrium long after the reaction begins appears to be limited by the rate of establishing fluctuations of a diffusion dependent concentration, therefore asymptotically at $t \to \infty$, instead of dependences of formal chemical kinetics

$$\overline{A}(t) = A(\infty) + fe^{-Wt}, \quad \overline{B}(t) = \overline{C}(t) = B(\infty) - fe^{-Wt}$$

the following law is obtained

$$A(t) = A$$

As before, averaging is performed over volume.

We shall consider the simplest case. Assume that at initial time $t = 0$ $A = A_0$, $B = C = 0$ and that in equilibrium $B(\infty) = C(\infty) \ll A(\infty)$, so that the change of A during the establishment of equilibrium $A(t) \approx A_0$, concentration fluctuations of initial substance A can be ignored.

Let us take a closer look at the fluctuations and divide them into correlated $q = \delta B + \delta C$ and anti-correlated ones $p = \delta B - \delta C$.

Separate decay acts of A molecules, as well as acts of B and C composition, are random events and thus create fluctuations. However, it is clear that in this event only correlated fluctuations are originated; the chemical reaction does not change the difference of concentrations. Only the diffusion of B and C in space may change value p.

Let us consider the rather late moment, when local thermodynamic equilibrium and equilibrium distribution $q(\infty)$ have been established everywhere, but value $p(t)$ connected with the slow process—diffusion has not yet achieved equilibrium. At this time period relationship

$$C(x, t)B(x, t) = B^2(\infty) \tag{3.123}$$

is valid.

Averaging (3.123) taking into account that

$$\overline{B} = B(\infty) + \frac{1}{2}p + \frac{1}{2}q$$

$$\overline{C} = C(\infty) + \frac{1}{2}p - \frac{1}{2}q, \qquad \overline{p} = \overline{q} = \overline{pq} = 0$$

one derives

$$\overline{B^2(t)} = B^2(\infty) - \frac{1}{4}\overline{q^2}(t) + \frac{1}{4}\overline{p^2}(t)$$

The problem is reduced to calculating the root-mean-square of the anti-correlated part of fluctuations. For estimation it should be noted that at time t fluctuations p with characteristic size $\sqrt{Dt}$ (where D is the diffusion coefficient) remained unexcited; the equilibrium amplitude of such fluctuations would be $\overline{p^2}(t) \approx B(\infty)/V$. Here the volume is $V \sim$ $\sim D^{3/2}t^{3/2}$; B is expressed in absolute units (the number of particles in cm^3). In equilibrium $q^2(\infty) = p^2(\infty)$, $\overline{B} = B(\infty)$, but at time t, because of partial excitation of anti-correlated fluctuations, we have

$$p^2(t) - q^2(t) \approx p^2(t) - p^2(\infty) \approx - B(\infty)/(D^{3/2}t^{3/2})$$

and finally

$$B(t) = B(\infty) - \alpha/(D^{3/2}t^{3/2}) \tag{3.124}$$

where α is a dimensionless multiplier of near unity. The given estimate can be easily made more accurate if we proceed to Fourier-representation $p(x, t) = \int p_K \exp(iKx) d^3K$. To determine p_K let us use a diffusion equation. The classical equation $\partial p/\partial t = D\Delta p$ would have given $dp_K/dt = K^2 D p_K$. However, in reality one should employ a diffusion equation with a random force describing fluctuations. Then the equation is written not for the deterministic p_K but for the probability distribution of systems with respect to p_K:

$$\Phi = \prod_K \Phi_K(p_K, t), \quad \frac{\partial \Phi}{\partial t} = - \frac{\partial}{\partial p_K} (K^2 D p_K \Phi_K) + D_K \frac{\partial^2 \Phi_K}{\partial p_K^2}$$

where D_K is the diffusion coefficient in p_K space.

Our objectives make it sufficient to derive an equation for the second moment:

$$\frac{d}{dt} \int_{-\infty}^{\infty} p_K^2 \Phi_K dp_K = -2K^2 D\overline{p_K^2} + 2D_K$$

Thus each $\overline{p_K^2}(t)$ tends to its stationary value in accordance with the exponential law. Stationary value $\overline{p_K^2}(\infty)$ is determined by the equilibrium spectrum, $p_K^2(\infty) = \text{const}\ (K)$ (white noise). Hence the same result (3.124) is obtained:

$$\overline{p^2}(x, t) = Q - \text{const}\ D^{-3/2}t^{-3/2} \tag{3.125}$$

Term $Q(Q \to \infty)$ is cancelled out with the same term in expression $q^2(x, t)$. And finally it should be noted that a change of asymptotics is also observed in the case of a decay into two equivalent molecules ($A \rightleftarrows 2B$), though the constant in (3.125) will differ by the absolute value.

Thus, law $t^{-3/2}$ is valid for all reactions in the course of approaching equilibrium. It is possible that the application of modern precise methods of investigation, in particular, light scattering by small angles, will permit tracing certain Fourier-components of the concentration, and not only its mean value. Taking into account A, B, C interaction may change the equilibrium constant and yield corrections to the mass action law. However, conclusions made on the kinetics of the process preserve their value. Long range Coulomb interactions at electrolytic dissociation $A \rightleftarrows B^+ + C^-$ constitute the only exception, but even here there is a large background of other ions not participating in the reaction. The condition of electroneutrality does not require the fulfilling of equality $B^+ = C^-$; consequently the propositions given above can be applied.

3.6.3. Role of Initial Nonequilibrium Density Fluctuations in the Kinetics of Diffusion Spreading

The role of initial fluctuations in reagent distribution may also appreciably influence kinetic regularities of chemical processes in the solid phase. The initial distribution of reagents under these conditions may be in a strong non-equilibrium state. Let us note that using the most general assumptions on the type of initial distributions and boundary conditions (see Chapter 4), it is extremely difficult to calculate the kinetics of a solid phase diffusion controlled process. We shall confine ourselves to considering one of the problems which is of interest because the method devised for solving it may be used in analyzing the kinetics of complicated solid phase chemical transformations.

Let us examine the spreading of two types of particles A and B enclosed in volume V. Initially (at time $t = 0$) they are localized in spheric granules. Each of the granules contained either A (letting N_1 represent the number of these granules in volume V) or B particles (the corresponding number of granules—N_2). Write $A(B)$ for the total number of particles $A(B)$, and $n_1 = A/V$ and $n_2 = B/V$ for the corresponding mean concentrations of particles. The process of diffusion spreading of particles over volume V may be stipulated, for example, by an instantaneous temperature rise of the whole system (in the limit by melting the solid). Our task is to describe the kinetics of such a diffusion spreading, or to be more exact, to search for the time dependence of function $\overline{Z}(r, t) = \overline{|p_1(r, t) - p_2(r, t)|}$. This time dependence represents the averaged over-the-whole volume absolute value of the difference of local concentrations $p_1(r, t)$ and $p_2(r, t)$ corresponding to A and B type particles (r is the radius-vector of particles). Evidently at large times

$$\overline{Z}(t) \longrightarrow |n_1 - n_2|$$

For simplicity we shall take the diffusion coefficients of particles A and B to be equal. In this case, the following diffusion equation is valid for function $Z(r, t) = \rho_A(r, t) - \rho_B(r, t)$:

$$\frac{\partial Z}{\partial t} = D\Delta Z \tag{3.126}$$

under additional conditions

$$Z(r, t) \longrightarrow 0 \qquad \partial Z/\partial r \longrightarrow 0$$

$$t \longrightarrow \infty$$

where $Z(r,0) = Z_0(r)$ is the initial value of function Z.

The solution of equation (3.126) can be presented in the form

$$Z(r, t) = \int G(r, r'; t) Z_0(r') dr' \tag{3.127}$$

where

$$G(r, r'; t) = \frac{1}{8(\pi D t)^{3/2}} \exp\left[-\frac{(r - r')^2}{4Dt}\right]$$

is the corresponding Green's function.

Let us write the expression for function $Z_0(r)$, in the explicit form taking into consideration the random character of the distribution of granules (spheres) of both type over the entire volume V:

$$\begin{aligned} Z_0(r) &= \frac{1}{2} \sum_{i=1}^{N} \eta(r - R_i)\left[\rho_1(1 + \sigma_{R_i} - \rho_2\left(1 - \sigma_{R_i}\right)\right] = \\ &= \frac{1}{2} \rho_1 \sum_{i=1}^{N} \eta(r - R_i)\left[(1 - \xi) + \sigma_{R_i}(1 + \xi)\right] \end{aligned} \tag{3.128}$$

Here

$$N = N_1 + N_2, \qquad \xi = \rho_2/\rho_1, \qquad \sigma_{R_i} = \pm 1$$

$$\eta(r - R_i) = \begin{cases} 1, & |r - R_i| < a_i \\ 0, & |r - R_i| > a_i \end{cases}$$

where R_i and a_i are the radius-vector and the radius of an i-th sphere.

It can be easily made certain that the integral over the whole volume of function $Z_0(r)$ according to the meaning of this function yields the difference of the full number of A and B particles in the entire volume V:

$$\begin{aligned} \int_V drZ_0(r) &= \frac{1}{2} \sum_{i=1}^{N} \nu_i \left[\left(1 + \sigma_{R_i}\right)\rho_1 - \rho_2\left(1 - \sigma_{R_i}\right)\right] \\ &= \sum_{i=1}^{N} \nu_i^{(1)} \rho_1 - \sum_{i=1}^{N} \nu_i^{(2)} \rho_2 = A - B \end{aligned}$$

where $\nu_i^{(1)}$ and $\nu_i^{(2)}$ are the corresponding volumes on an i sphere, if it contains A or B particles.

In further discussion, our interest will lie with the tempral change of absolute value $|Z(r,t)|$. To make the following calculations more convenient, we shall use integral representation:

$$|Z(r,t)| = \frac{2}{\pi}\int_0^\infty \xi^{-2}(1 - \cos \xi Z(r,t))d\xi$$

which upon substituting $Z_0(r)$ from (3.127) into (3.128) and averaging over the position of spheres in space gives

$$\langle |Z(r,t)| \rangle = -\frac{2}{\pi}\int_0^\infty \frac{d\xi}{\xi^2}\left\{\left\langle \cos\left(\frac{1}{2}\xi\rho_1 \sum_{i=1}^N \left[(1-\xi) + \right.\right.\right.\right.$$

$$\left.\left.\left.\left. + \sigma_{R_i}(1+\xi)\right]\right)\tilde{G}(r, R_i;\ t)\right\rangle - 1\right\}$$

where

$$\tilde{G}(r,R;\ t) \equiv \tilde{G}_i = \int d\rho G(r, \rho + R_i;\ t)\eta(\rho) =$$

$$= \int_{\nu_i} d\rho G(r, \rho + R_i;\ t)$$

Carrying out the corresponding calculations one obtains

$$\langle |Z(r,t)| \rangle = \frac{2}{\pi}\int_0^\infty \frac{d\xi}{\xi^2}\left(1 - \frac{1}{2}\prod_{i=1}^N \left\langle \exp\left(\frac{i}{2}\xi\rho_1\tilde{G}_i\left[(1-\xi) + \right.\right.\right.\right.$$

$$\left.\left.\left. + \sigma_{R_i}(1+\xi)\right]\right)\right\rangle - \frac{1}{2}\prod_{i=1}^N \left\langle \exp\left(-\frac{i}{2}\xi\rho_1\tilde{G}_i\left[(1-\xi) + \right.\right.\right.$$

$$\left.\left.\left.\left. + \sigma_{R_i}(1+\xi)\right]\right)\right\rangle\right) = \frac{2}{\pi}\int_0^\infty \frac{d\xi}{\xi^2}\left(1 - \frac{1}{2}\prod_{i=1}^N \times\right.$$

$$\times [\lambda_1 \exp(i\xi\rho_1\tilde{G}_i) + \lambda_2 \exp(-i\xi\rho_2\tilde{G}_i)] - \frac{1}{2}\prod_{i=1}^{N} \times$$

$$\times [\lambda_1 \exp(-i\xi\rho_1\tilde{G}_i) + \lambda_2 \exp(i\xi\rho_2\tilde{G}_i)]\Bigg\} \qquad (3.129)$$

where

$$\lambda_i = \frac{N_i}{N_1 + N_2} = \frac{N_i}{N}\ (i = 1,2), \qquad \rho_2 = \xi\rho_1$$

Evidently, after being averaged over volume V values $\tilde{G}_i$ will have small parameter v_i/V. Hence let us expand the exponentials in (3.129) as series preserving only terms up to second order in $\tilde{G}_i$ and then average the expression over R_i (shown by a line above the symbol). The result is

$$\langle|\bar{Z}|\rangle = \frac{2}{\pi}\int_0^\infty \frac{d\xi}{\xi^2}\Bigg\{1 - \frac{1}{2}\Bigg[1 + i\xi(\lambda_1\rho_1 - \lambda_2\rho_2)\sum_{i=1}^{N}\overline{\tilde{G}_i} -$$

$$-\frac{1}{2}\xi^2(\lambda_1\rho_1^2 + \lambda_2\rho_2^2)\sum\overline{\tilde{G}_i^2} + \ldots\Bigg] - \frac{1}{2}i\xi(\lambda_1\rho_1 - \lambda_2\rho_2)\times$$

$$\times\sum_{i=1}^{N}\overline{\tilde{G}_i} - \frac{1}{2}\xi^2\sum\overline{\tilde{G}_i^2}(\lambda_1\rho_1^2 + \lambda_2\rho_2^2) + \ldots\Bigg]\Bigg\} =$$

$$= \frac{2}{\pi}\int_0^\infty d\xi\,\xi^{-2}[1 - \exp(-\lambda\xi^2)\cos\xi\mu] \qquad (3.130)$$

Here

$$\lambda = \frac{1}{2}\sum_{i=1}^{N=N_1+N_2}(\lambda_1\rho_1^2 + \lambda_2\rho_2^2)\,\overline{\tilde{G}_i}\,;$$

$$\mu = \sum_{i=1}^{N}(\lambda_1\rho_1 - \lambda_2\rho_2)\,\overline{\tilde{G}_i}$$

Taking into consideration that

$$\bar{\tilde{G}}_i = \frac{\nu_i}{V}, \qquad \bar{\tilde{G}}_i^{\,2} = \frac{\sqrt{2\pi}\, a_i^6}{18V(Dt)^{3/2}}$$

it can be easily found that

$$\mu = V^{-1}\left(\sum_{i=1}^{N_1} \rho_1 \nu_i^{(1)} - \sum_{i=1}^{N_2} \rho_2 \nu_i^{(2)}\right) = n_1 - n_2$$

$$\lambda = \frac{1}{16(2nDt)^{3/2} V}\left[\sum_{i=1}^{N_1} \rho_1^2 (\nu_i^{(1)})^2 + \sum_{i=1}^{N_2} \rho_2^2 (\nu_i^{(2)})^2\right] \tag{3.131}$$

Integrating (3.154) one can derive

$$\langle|\bar{Z}|\rangle = |\mu|\left[1 + \frac{1}{2}\Gamma\left(-\frac{1}{2}, \frac{\mu^2}{4\lambda}\right)\right] \tag{3.132}$$

where $\Gamma(a, x)$ is an incomplete gamma function [200]. Within the limit of large times at $\mu^2/4\lambda \gg 1$ we obtained from (3.132):

$$\langle|\bar{Z}|\rangle = |\mu|\left[1 + 4\frac{\lambda^{3/2}}{|\mu|^3}\exp\left(-\frac{\mu^2}{4\lambda}\right)\right] \tag{1.133}$$

It becomes clear from (3.133) that at $t \to \infty$ $(\lambda \to 0)$ value $<|\bar{Z}|> \to$ $\to n_1 - n_2$ as could have been expected.

Expressions (3.132) and (3.133) are valid if the mean concentrations of reagents are unequal, i.e., if $\mu \neq 0$. If $n_1 = n_2$, i.e., $\mu = 0$ it follows from (3.154) that

$$\langle|\bar{Z}|\rangle \sim t^{-3/4}$$

In the general case at $\mu \neq 0$ it is convenient to introduce value α, characterizing the degree of diffusion "spreading" of the substance. In particular, at $n_2 < n_1$

$$\alpha = n_2^{-1}\left[\langle|\bar{Z}|\rangle - \langle|\bar{Z}|\rangle_\infty\right] =$$

$$= 1 - \frac{1}{n_2}\,\frac{4\lambda^{3/2}}{(n_1 - n_2)^2}\exp\left[-\frac{(n_1 - n_2)^2}{4\lambda}\right] \tag{3.134}$$

Substituting explicit expression (3.131) for λ into (3.134), we obtain the characteristic dependence with $n = 3/2$.

Such a result for the kinetics of spreading (a faster establishment of equilibrium than in the case of a first order chemical reaction) is not surprising since in the considered problem the whole number of particles does not decrease (as in the case of a chemical reaction) but remains constant, so that all the particles participate in establishing the final thermodynamic state.

It is obvious that taking into account the "chemical activity" upon the collision of diffusing particles (particles escape the ensemble of diffusing particles upon collision) should lead to a change of kinetic dependences.

3.7. THEORETICAL—FIELD DESCRIPTION OF THE DIFFUSION KINETICS OF REACTING PARTICLES

In the previous sections, which discussed the consideration of fluctuations in DCR, we treated asymptotic laws of changes in particle concentration in the course of the reaction. The methods applied were nonrigorous, heuristic. Recently more consistent methods have been developed in DCR kinetics for an ensemble of particles considering the many-particle character of the problem. They are based on an exact expansion of a solution for a many-particle distribution function in low particle concentrations. To consider the problem in this way it is convenient to use operator equations for many-particle distribution functions, the same as in field theory and statistical physics [12].

3.7.1. Operator Expression of Basic Equations

Let us consider an example of a chemical reversible reaction $B + B \rightleftarrows A$. We shall start by deriving the corresponding operator equations from the simplest case—when the system consists of two B particles. Probability density for the first particle to be located in point x_1 and the second—in point x_2—will be assigned by function $u_{2,0}(x_1,x_2)$, while the probability density for the created one particle A to be located in point y will be established by function $u_{0,1}(y)$. For the given problem of two particles, the condition of normalizing to unity of the full probability for being the system in some state has the form

$$\int dx_1 dx_2 u_{2,0}(x_1,x_2) + \int dy u_{0,1}(y) = 1 \tag{3.135}$$

Let function $w_-(x_1 - x_2)$ prescribe the probability for two B particles (in points x_1 and x_2) to transform into an A particle. The probability for an A particle located in point y to transform into two B particles, one of which is in y the other in x, is defined by $w_+(y - x)$. There is no need to distinguish B particles, therefore function $u_{2,0}(x_1,x_2)$ may be considered a symmetric function of its arguments.

Now the equation for the evolution of this system can be written:

$$\frac{\partial u_{2,0}(x_1,x_2)}{\partial t} = D_b\left(\Delta_{x_1} + \Delta_{x_2}\right)u_{2,0}(x_1,x_2) -$$

$$- 2w_-(x_1 - x_2)u_{2,0}(x_1,x_2) + w_+(x_1 - x_2)u_{0,1}(x_2) +$$

$$+ w_+(x_2 - x_1)u_{0,1}(x_1) \qquad (3.136a)$$

$$\partial u_{0,1}(x)/\partial t = D_a \Delta_x u_{0,1}(x) + 2\int w_-(x_1 - x_2)u_{2,0}(x_1,x_2)dx_2 -$$

$$- 2\int w_+(x_1 - x_2)dx_2 u_{0,1}(x_1) \qquad (3.136b)$$

Terms for the creation and annihilation of *A* and *B* particles in the right parts of Equations (3.136) are quite evident from the definition of functions $w_+(x)$ and $w_-(x)$ (D_a and D_b are diffusion coefficients of *A* and *B* particles respectively). The condition of conserving full probability (3.135) readily follows from (3.136). To generalize equations of type (3.136) to a many-particle system, it is necessary to introduce functions

$$u_{N,M}(x_1,x_2,\ldots,x_N | y_1,y_2,\ldots,y_M)$$

symmetrically separate in coordinates x_i and y_i. Avoiding writing the corresponding bulky system of equations in x-representation, let us note while looking at the system of Equations (3.136) that it somewhat resembles a system of equations of quantum field theory for the creation and annihilation of particles in the Fock representation [192]. Also considering the symmetry in variables x and y separately, one can try to present this system using the operators of creation and annihilation of these particles $\Psi_a^+(x)$, $\Psi_a(x)$ and $\Psi_b^+(x)$, $\Psi_b(x)$ which possess commutation relationships of Bose operators

$$[\Psi_a(x), \Psi_a^+(y)] = \delta(x - y), \qquad [\Psi_a^+(x), \Psi_a^+(y)] = 0$$

$$[\Psi_a(x), \Psi_a(y)] = 0$$

Analogous relationships are chosen for *B* particles. Operators of *A* and *B* particles commutate.

A new representation of the system (3.136) appears to be possible if an operator function is introduced

$$\Phi = \sum_{M,N=0} \int\ldots\int dx_1\ldots dx_N dy_1\ldots dy_M u_{N,M}(x_1,\ldots,x_N | y_1,\ldots,y_M) \times$$

$$\times \Psi_b^+(x_1)\Psi_b^+(x_2)\ldots\Psi_b^+(x_N)\Psi_a^+(y_1)\Psi_a^+(y_2)\ldots\Psi_a^+(y_M)$$

and the "Hamiltonian" of the system

$$\hat{\mathscr{H}} = -D_b \int dx \nabla_x \Psi_b^+(x) \nabla_x \Psi_b(x) - D_a \int dx \nabla_x \Psi_a^+(x) \nabla_x \Psi_a(x) +$$

$$+ \frac{1}{2} \iiint dx_1 dx_2 dy \, [\delta(y - x_1) + \delta(y - x_2)] w_+(x_1 - x_2) \times$$

$$\times \Psi_b^+(x_1) \Psi_b^+(x_2) \Psi_a(y) + \frac{1}{2} \iiint dx_1 dx_2 dy \, [\delta(y - x_1) +$$

$$+ \delta(y - x_2)] w_-(x_1 - x_2) \Psi_b(x_1) \Psi_b(x_2) \Psi_a^+(y) -$$

$$- \iint dx_1 dx_2 w_-(x_1 - x_2) \Psi_b^+(x_1) \Psi_b^+(x_2) \Psi_b(x_2) \Psi_b(x_1) -$$

$$- \iint dx dx' w_+(x - x') \Psi_a^+(x) \Psi_a(x)$$

Then the evolution equation can be written in the form

$$\frac{\partial \Phi}{\partial t} = \hat{\mathscr{H}} \Phi$$

i.e., analogously to the Schroedinger equation.

In spite of a formal resemblance they still differ significantly. Operator $\hat{\mathscr{H}}$ for example is not Hermitian. But more important is that the determination of physical values (a mean one) using operator Φ is carried out according to absolutely different rules. To set up a norm

$$\sum_{N, M=0}^{\infty} \int \dots \int dx_1 \dots dx_N dy_1 \dots dy_M u_{N, M}(x_1, \dots, x_N | y_1, \dots, y_M)$$

with respect to operator Φ one should take the usual theoretical field scalar product of Φ into function Φ_0 defined in the following way

$$\langle \Phi_0 | = \langle 0 | \sum_{K, K'=0}^{\infty} (K!)^{-1} \left[\int \Psi_b(x) dx \right]^K (K')^{-1} \left[\int \Psi_a(x) dx \right]^{K'} =$$

$$= \langle 0 | \exp \left\{ \int \left[\Psi_a(x) + \Psi_b(x) \right] dx \right.$$

The normalization condition can be presented in the form of equality

$$\langle \Phi_0 | \Phi \rangle = 1$$

It is actually satisfied which follows from relation

$$\langle\Phi_0|\hat{\mathscr{H}} \equiv 0$$

The number of particles, say A, is expressed as the mean one from operator

$$\hat{N}_A = \int \Psi_a^+(x)\Psi_a(x)\,dx$$

namely

$$N_A(t) = \langle\Phi_0|\,\hat{N}_A\,|\Phi\rangle \tag{3.137}$$

There is also a Heisenberg representation of operators, since

$$|\Phi(t)\rangle = \exp(\hat{\mathscr{H}}t)\,|\Phi_{in}\rangle$$

(where $|\Phi_{in}\rangle$ is the initial function of the system) and the average of type (3.137) can be written in the following way

$$N_A(t) = \langle\Phi_0|\,\exp(-\hat{\mathscr{H}}t)\hat{N}_A\,\exp(\hat{\mathscr{H}}t)\,|\Phi_{in}\rangle$$

Thus operator $\hat{N}_A(t)$ in a Heisenberg representation has the form

$$\hat{N}_A(t) = \exp(-\hat{\mathscr{H}}t)\hat{N}_A\,\exp(\hat{\mathscr{H}}t)$$

and obeys the usual equations of motion

$$d\hat{N}_A/dt = [\hat{N}_A, \hat{\mathscr{H}}]$$

So motion equations of the classical system can be presented in the quantum form. Note that the idea of expressing the classical Liouville equation in mechanics in the quantum form was first suggested by Prigogine [193].

It is convenient to express Hamiltonian $\hat{\mathscr{H}}$ using operators of creation in the momentum representation as is accepted in statistical mechanics [194]:

$$a_K^+ = \frac{1}{\sqrt{V}} \int \Psi_a^+(x)e^{iKx}\,dx, \qquad \beta_K^+ = \frac{1}{\sqrt{V}} \int \Psi_b^+(x)e^{iKx}\,dx$$

Then we have

$$\hat{\mathscr{H}} = -\sum_{K} D_b K^2 \beta_K^+ \beta_K - \sum_{K} D_a K^2 a_K^+ a_K +$$

$$+ \frac{1}{2\sqrt{V}} \sum_{K_1+K_2=9} [w_+(K_1 - 9) + w_+(K_2 - 9)] \beta_{K_1}^+ \beta_{K_2}^+ \alpha_q +$$

$$+ \frac{1}{2\sqrt{V}} \sum_{K_1+K_2=9} [w_-(K_1 - 9) + w_-(K_2 - 9)] \beta_{K_1} \beta_{K_2} \alpha_q^+ +$$

$$+ \frac{1}{V} \sum_{K_1+K_2=K_1'+K_2'} \beta_{K_1}^+ \beta_{K_2}^+ \beta_{K_2'} \beta_{K_1'} w_-(K_1 - K_1') -$$

$$- \sum_{K} w_+^0 \alpha_K^+ \alpha_{K'} \tag{3.138}$$

where $w_+(K)$ and $w_-(K)$ are Fourier transforms of functions $w_+(x)$ and $w_-(x)$; w_+^0 is the value of $w_+(K)$ at $K=0$.

3.7.2. Equations of Particle Density Evolution and Spectrum of Fluctuations in the Chemical Reaction Process

Hamiltonian (3.138) resembles the Hamiltonian of an interacting Bose gas (though non-Hermitian). The methods of determining its eigenfunctions are adopted from Bogolyubov's theory of a weakly non-ideal Bose gas [195]. The weakness of the interaction is guaranteed by low particle density, i.e., by parameter $n_A r_A^3 \ll 1$ (where r_A is the radius of A particles). To avoid needlessly complicated calculations we assume functions $w_+(K)$ and $w_-(K)$ reduced to a dimensionless form and to be small*.

Let us make use of the fact that in the case of low density there is a macroscopic number of A and B particles with zero momentum and, consequently, operators $\alpha_0^+, \beta_0^+, \alpha_0, \beta_0$ may be regarded as c-numbers. Expanding the Hamiltonian according to Bogolyubov's theory with an accuracy up to quadratic terms in $\alpha_K, \beta_K, \beta_K^+, \alpha_K (K \neq 0)$ one can present $\hat{\mathscr{H}}$ as a sum of two terms: $\hat{\mathscr{H}} = \hat{\mathscr{H}}_0 + \hat{\mathscr{H}}_1$ where

---

*Otherwise it is necessary to make an extra summing of ladder diagrams of the perturbation theory to take into consideration the dynamics of the motion with potential $w_-(x)$. Our approximation is equivalent to Born's one in quantum mechanics.

$$\hat{\mathcal{H}}_0 = \frac{1}{\sqrt{V}}\, w_+^0 (\beta_0^+)^2 \alpha_0 + \frac{1}{\sqrt{V}}\, w_-^0 \beta_0^2 \alpha_0^+ -$$

$$- \frac{1}{V}\, w_-^0 (\beta_0^+)^2 \beta_0^2 - w_+^0 \alpha_0^+ \alpha_0$$

and $\hat{\mathcal{H}}_1$ is the square in $\beta_K, \alpha_K, \alpha_K^+, \beta_K^+ (K \neq 0)$ part of Hamiltonian $\hat{\mathcal{H}}$ which will be given below. Since α_0 and β_0 are essentially classical variables, the motion equations can be obtained for them by writing motion equations as for Heisenberg operators and then considering them as c-numbers. We obtain at this

$$\frac{d\beta_0}{dt} = 2\,\frac{w_+^0}{\sqrt{V}}\, \alpha_0 \beta_0^+ - \frac{2w_-^0}{V}\, \beta_0^+ \beta_0^2 -$$

$$- \frac{2}{V}\, \beta_0^+ \left\langle \sum_K w_-(K)\, \beta_K(t)\, \beta_{-K}(t) \right\rangle \tag{3.139}$$

$$\frac{d\alpha_0}{dt} = \frac{1}{\sqrt{V}}\, w_-^0 \beta_0^2 - w_+^0 \alpha_0 + \frac{1}{\sqrt{V}} \left\langle \sum_K w_-(K)\beta_K(t)\, \beta_{-K}(t) \right\rangle$$

The equations are written omitting some terms which later will be shown to equal zero. The equations for α_0^+ and β_0^+ appear to be analogous and are valid at

$$\alpha_0^+ = \beta_0^+ = \sqrt{V} \tag{3.140}$$

Let us introduce densities $a = \alpha_0(\sqrt{V})^{-1}$ and $b = \beta_0(\sqrt{V})^{-1}$ of A and B particles respectively* instead of α_0, β_0 and taking into consideration (3.140) Hamiltonian $\hat{\mathcal{H}}_1$ in the following way:

$$\hat{\mathcal{H}}_1 = - \sum_K \varepsilon_1(K)\beta_K^+ \beta_K - \sum_K \varepsilon_2(K)\alpha_K^+ \alpha_K +$$

$$+ \sum_K \mu_+(K)\beta_K^+ \alpha_K + \sum_K \mu_-(K)\alpha_K^+ \beta_K + \sum_K \lambda(K)\beta_K^+ \beta_{-K}^+$$

---

*Later it will be seen that $a = V^{-1} N_A(t) = \left\langle \Phi_0 \right| \sum_K a_K^+ a_K \left| \Phi \right\rangle =$

$= \alpha_0 V^{-1/2}, \quad V^{-1} N_B(t) = b(t) = \beta_0 V^{-1/2}$

where notations are introduced

$$\varepsilon_1(K) = D_b K^2 + 2b[w_-^0 + w_-(K)], \qquad \mu_+(K) = w_+^0 + w_+(K)$$

$$\varepsilon_2(K) = D_0 K^2 + w_+^0, \qquad \mu_-(K) = w_-^0 + w_-(K)$$

$$\lambda(K) = aw_+(K) - b^2 w_-(K)$$

Motion equations with Hamiltonian $\hat{\mathscr{H}}_1$ for mean values

$$\sigma_{\beta\beta}(K, t) = \langle 0|\beta_K(t)\,\beta_{-K}(t)\,|\Phi_{in}\rangle$$

$$\sigma_{\alpha\alpha}(K, t) = \langle 0|\alpha_K(t)\,\alpha_{-K}(t)\,|\Phi_{in}\rangle$$

$$\sigma_{\alpha\beta}(K, t) = \langle 0|\alpha_K(t)\,\beta_{-K}(t)\,|\Phi_{in}\rangle = \sigma_{\beta\alpha}(-K, t)$$

have the form

$$\frac{d\sigma_{\beta\beta}(K, t)}{\partial t} = -2\varepsilon_1(K)\sigma_{\beta\beta}(K, t) + 2\mu_+(K)\sigma_{\beta\alpha}(K, t) + 2\lambda(K)$$

$$\frac{d\sigma_{\alpha\alpha}(K, t)}{\partial t} = -2\varepsilon_2(K)\sigma_{\alpha\alpha}(K, t) + 2\mu_-(K)\sigma_{\beta\alpha}(K, t) \tag{3.141}$$

$$\frac{d\sigma_{\alpha\beta}(K, t)}{\partial t} = -[\varepsilon_1(K) + \varepsilon_2(K)]\sigma_{\alpha\beta}(K, t) + \mu_+(K)\sigma_{\alpha\alpha}(K, t) + \mu_-(K)\sigma_{\beta\beta}(K, t)$$

If we add the first of Equations (3.139) written in the same notations:

$$\frac{db}{dt} = 2w_+^0 a - 2w_-^0 b^2 - \frac{2}{V}\sum_K w_-(K)\sigma_{\beta\beta}(K, t) \tag{3.142}$$

and the conservation of the number of particles

$$b + 2a = \text{const} \tag{3.143}$$

which follows from the second equation of (3.139) we obtain the sought after system of equations. It simultaneously defines the dynamics of changes in the number of particles resulting from the course of the chemical reaction and the dynamics of the development of fluctuations in time. Note that though system of equations for $\sigma_{\alpha\alpha}, \sigma_{\alpha\beta}, \sigma_{\beta\beta}$ is linear, its coefficients ε_1 and $\lambda(K)$ are time dependent through functions $b(t)$ and $a(t)$.

3.7.3. Corrections to the Mass Action Law

To determine the corrections to the Mass Action Law, one should consider an equilibrium situation, i.e., time independent $a(t)$, $b(t)$, and $\sigma_{\alpha\alpha}$, $\sigma_{\alpha\beta}$, $\sigma_{\beta\beta}$. First we determine (3.141) and then substitute it into (3.142). Neglecting fluctuations (i.e., $\sigma_{\beta\beta}$) in Equation (3.142) gives the mass action law in its classical form

$$b^2/a = w_+^0/w_-^0 = K_p(T)$$

Adding (3.143) one determines equilibrium concentrations a_0 and b_0. The consideration of fluctuation terms leads to a correction to equilibrium concentrations: $a_{eq} = a_0 + \delta a$, where

$$\delta a = -\frac{\delta b}{2} = w_0^{-1} V^{-1} \sum_K w_-(K)\lambda(K) \times$$

$$\times \frac{\varepsilon_1(K)\varepsilon_2(K) - \mu_+(K)\mu_-(K) + \varepsilon_2^2(K)}{[\varepsilon_1(K)\varepsilon_2(K) - \mu_+(K)\mu_-(K)][\varepsilon_1(K) + \varepsilon_2(K)]};$$

$$w_0 = w_+^0 + 4b_0 w_-^0 \tag{3.144}$$

Here we shall not take the integrals for any concrete functions $w_+(K)$ and $w_-(K)$, as we are mainly interested in the fact that such corrections exist. Note that generally speaking density corrections to the mass action law can be of two types. The first type are potential corrections stipulated by the fact that reacting particles are connected by an interaction potential. Corrections of the second type (let us call them kinetic) are connected with the fact that with a certain probability particles may react. These kinetic corrections are given by formula (3.144). Certainly the principle of detailed equilibrium for $w_\pm(K)$ ensures the coincidence of the final result with the equilibrium one calculated on the basis of thermodynamic concentrations.

3.7.4. Asymptotic Behavior of Densities $a(t)$ and $b(t)$

To determine the characteristic asymptotic time dependences of densities $a(t)$ and $b(t)$, we shall consider a simple non-stationary problem. Let the system of A and B particles at first be in a state of equilibrium at temperature T. The densities of particles a_0 and b_0 are described as in Section 3.7.3. Let us assume that we suddenly increase the temperature by δT. Such characteristics of the system as the diffusion coefficients, probabilities of dissociation and recombination $w_+(K)$, $w_-(K)$ will also suddenly change. According to the basic system of equations, equilibrium concentrations of A and B particles should also change. Let us consider the approaching of $a(t)$ and $b(t)$ to their new values.

Since initial and final densities do not differ much (because of the small increment in temperature) the changes of $\varepsilon_{1,2}, \mu_\pm$ in time can be neglected and the entire system of Equations (3.141–3.143) becomes linear with constant coefficients.

Let us, for example, consider the general solution of the linear system of Equations (3.141) without the term with $\lambda(K)$. In this case $\sigma_{\beta\beta}(K, t)$ will contain time exponents of the type

$$\exp[-2w_1(K)t], \qquad \exp\{-[w_1(K) + w_2(K)]t\} \qquad \exp[-2w_2(K)t]$$

where

$$w_{1,2}(K) = \frac{\varepsilon_1(K) + \varepsilon_2(K)}{2} \pm \left\{\frac{[\varepsilon_1(K) - \varepsilon_2(K)]^2}{4} + \mu_+(K)\mu_-(K)\right\}^{1/2}$$

The asymptotics at $t \to \infty$ of value

$$\frac{1}{V}\sum_K \sigma_{\beta\beta}(K, t)w_-(K)$$

in (3.142) is determined by the small K. At small K the value $w_1(K)$ tends to w_0, i.e., to the finite value, whereas $w_2(K)$ tends to $D_{\mathrm{eff}}K^2$ and yields a power time dependence of the above value and, together with it, of particle density $a(t)$.

Indeed, at $w_0t \gg 1$ the terms in $\sigma_{\beta\beta}(K, t)$ containing exponent $\exp[-w_1(K)t]$ can be neglected. Finding the solution from (3.141), one obtains for $\sigma_{\beta\beta}(K, t)$

$$\sigma_{\beta\beta}(K, t) \longrightarrow \tilde{\sigma}_{\beta\beta}(K, t) = \frac{1}{(w_1 - w_2)^2}\,[\mu_+^2(K)\,\delta\sigma_{\alpha\alpha}^{(0)}(K) +$$

$$+ (\varepsilon_1 - w_1)^2\delta\sigma_{\beta\beta}(K) - 2\mu_+(K)(\varepsilon_1 - w)\,\delta\sigma_{\alpha\beta}^{(0)}(K)] \times$$

$$\times \exp[-2w_2(K)t]$$

where $\delta\sigma_{\alpha\alpha}^{(0)}$, $\delta\sigma_{\alpha\beta}^{(0)}$ and $\delta\sigma_{\beta\beta}^{(0)}$ are the changes in the corresponding values in equilibrium due to a temperature increase by δT.

If

$$\frac{1}{V}\sum_K w_-(K)\tilde{\sigma}_{\beta\beta}(K, t)$$

changes according to the power-like time dependence, then at large t the term with derivative db/dt can be ruled out from (3.142) and the asymptotics at large t takes the form:

$$\delta a(t) \approx V^{-1} w_0^{-1} \sum_K w_-(K) \tilde{\sigma}_{\beta\beta}(K, t) = -\delta b(t)/2$$

To calculate the integral over K at $t \to \infty$, $K = 0$ must be taken everywhere in the pre-exponent, after which we get

$$\delta a(t) = \frac{w_+^0}{8 w_0^3 (2\pi D_{\text{eff}} t)^{3/2}} \left(\frac{\delta x}{\delta T}\right) \delta T$$

Here D_{eff} is the effective diffusion coefficient, which is expressed by initial parameters in the following way:

$$D_{\text{eff}} = \frac{w_+^0 D_b + 4bw_-^0 D_a - 2bw_-^0 w_+'}{w_+^0 + 4bw_-^0}, \qquad w_\pm' = \left.\frac{dw_\pm(K)}{dK^2}\right|_{K=0}$$

For x we have

$$x = (w_+' w_-^0 - w_+^0 w_-')a/w_-^0 D_{\text{eff}}$$

It is interesting to note that the effective diffusion coefficient does not go to zero, even if both diffusion coefficients D_a and D_b are equal to zero. The obtained diffusion coefficient is connected with the fact that the process of creation and annihilation in points other than the initial positions causes diffusion and equalization of concentration. Note that the exponent in asymptotics coincides with what has been obtained on the basis of heuristic considerations.

In conclusion, we shall mark that function $|\Phi\rangle$ which is the solution of equation

$$\partial|\Phi\rangle/\partial t = \hat{\mathscr{H}}|\Phi\rangle$$

has the structure

$$|\Phi\rangle = \exp\left\{\sum_K \left[\alpha_K^+ \alpha_{-K}^+ \gamma_{\alpha\alpha}(K, t) + \alpha_K^+ \beta_{-K}^+ \gamma_{\alpha\beta}(K, t) + \beta_K^+ \beta_{-K}^+ \gamma_{\beta\beta}(K, t)\right]\right\}$$

and consequently all mean values of type

$$<0|\alpha_K^{+}\alpha_K|\Phi>, \qquad <0|\beta_K^{+}\alpha_K|\Phi>, \qquad <0|\beta_K^{+}\beta_K|\Phi>,$$

$$<0|\alpha_K^{+}\beta_K|\Phi>, \qquad <0|\alpha_K^{+}\alpha_{-K}^{+}|\Phi>$$

are equal to zero which justifies dropping the mean ones (mentioned in Section 3.82) in deriving the basic equation.

CHAPTER 4

BOUNDARY CONDITIONS IN THE KINETICS OF DIFFUSION CONTROLLED REACTIONS

DCR theory reduces the problem of jointly describing the transfer processes and the chemical reaction to deriving transfer equations and adequately formulating boundary conditions that take into consideration the course of the chemical reaction [196,197].

Here are several principal propositions.

1. In diffusion kinetics, transfer processes are considered within the framework of nonequilibrium thermodynamics and described by equations containing phenomenological parameters as transfer coefficients. These equations are of a macroscopic character and describe the transport of reagents in bulk phases. Boundaries are neglected when deriving the equations and boundary conditions are formulated on the basis of simple physical considerations of the behavior of the corresponding values (concentration, temperature) on the boundaries of bulk phases.

However, the transfer equations are approximate and valid only for systems in a weakly non-equilibrium state (i.e., in a state close to complete thermodynamic equilibrium) when thermodynamic forces (e.g., gradient of concentrations or temperatures) and flows of the corresponding values are small. Describing systems characterized by a greater degree of non-uniformity (e.g., of chemical reactions) as well as revealing the physical sense of phenomenological transfer coefficients, requires proceeding to more general equations which can be obtained using methods of statistical mechanics. But the usually employed boundary conditions, having the sense of balance equations when passing through interphase boundaries, may appear inapplicable for equations of a more general type [196].

Boundaries do not influence the form of differential equations. However, the boundary conditions that must be satisfied by the equation solutions are not intuitive and can be determined only upon a detailed investigation of a more general equation of system evolution (master equation). The role of such general equations can be performed by difference equations of the random process theory [196] which we shall use in the analysis of boundary conditions for diffusion controlled reactions.

2. In the general case, the course of heterogeneous chemical reactions is connected with a transfer through or along the interphase boundary of mass, energy and momentum. Hence, in DCR kinetics the boundary should sometimes be treated as an independent phase and described on the basis of general methods of non-equilibrium thermodynamics [197] or statistical mechanics [198–199]. Such "surface" phases may have independent dynamics connected with the bulk phase dynamics. In the simplest cases, equations of surface phase dynamics reduce to a system of boundary conditions for equations of bulk phase dynamics [197]. These propositions are most significant for discussing the problem of boundary conditions in DCR kinetics. They prove that an adequate description requires revealing various scale factors of physical phenomena. Thus, the auxiliary problem in the Smoluchowski Method is solved by macroscopic formulation, and its results are used for considering the kinetics of chemical reactions at the molecular level.

Proceeding to a molecular-kinetic description of diffusion controlled chemical reactions also requires modifying the interpretation of the interphase boundary since it is necessary to consider the process on the reaction surface for a separate particle. In this case, most fruitful is the introduction of the notion of a boundary (or reaction) layer which divides two reaction regions arising in the course of chemical transformations proper [200–206].

Let us discuss the physical grounds for the notion of a boundary layer.

1. In traditional DCR theory (see Chapter 1), the mechanism of the process comprises two stages: a diffusion transfer of reaction-capable components which stipulates the possibility of their collision and an irreversible chemical reaction proceeding during the collision. The term collision encloses information on the duration of the reaction and the distances at which such a reaction is possible. Thus it is supposed that there are two regions: the region of purely diffusion motion (e.g., for a pair of particles) and the region of the reaction. These regions can be called outer (diffusional) and inner (of the reaction region). However, the dynamics of chemical reactions are usually described in terms of multidimensional configurational space (defined by the coordinates of all molecule atoms participating in the reaction). This imposes certain requirements on the description of processes in the outer region.

2. In the description of transfer processes, the solvent (or the medium in which the diffusion and reaction proceed) is treated as a structureless continuum [207] which determines the mobility of the dissolved reaction capable components. For describing only transport processes, such a representation appears to be sufficient, but it is too general to reveal the relation between diffusion and reaction, i.e., in the analysis of boundary regions. Hence the notions of diffusion mobility and interaction between dissolved particles and the solvent need detailing, which may appear quite difficult to do in the case of condensed media.

3. A general DCR is treated as a two-stage reaction: the diffusion stage with rate constant k_D and the chemical transformation stage with rate constant k. As a rule a DCR is considered irreversible, since the second stage is assumed to be irreversible. However it is unclear whether the first stage is reversible or not. Moreover, it is unclear what may be

considered a reverse reaction for the whole DCR. This is an important question since there are arguments pointing to a seeming violation of the principle of detailed balance in diffusion kinetics processes (see Section 1.3).

The second aspect of the problem lies with revealing the dependences of k_D and k on molecular parameters which is necessary for elucidating the sense of the phenomenological transfer coefficients of the boundary conditions at the molecular level.

4.1. DESCRIPTION OF DIFFUSION CONTROLLED REACTIONS WITHIN THE FRAMEWORK OF FORMAL KINETICS

For solving the above-mentioned problems consistently, let us consider DCR description within the framework of classical kinetics. And so, first of all, it is necessary to understand the meaning of the known [see (1.22)] addition Law of Kinetic Resistances:

$$k_{ex}^{-1} = k_D^{-1} + k^{-1}$$

Since the considered process includes two stages (its rate goes to zero if the rate constant in one of the successive stages is equal to zero) DCR can be modeled by a system of two successive reactions

$$A \underset{k_2}{\overset{k_1}{\rightleftarrows}} \mathrm{B} \underset{k_4}{\overset{k_3}{\rightleftarrows}} C \qquad (4.1)$$

In further discussion the stage of transition $B \rightleftarrows C$ is considered diffusional defining C as a mobile component. In formal kinetics for systems of successive reversible reactions (4.1), the following equations are valid:

$$dA/dt = -k_1 A + k_2 B$$

$$dB/dt = -(k_3 + k_2)B + k_1 A + k_4 C \qquad (4.2)$$

$$dC/dt = -k_4 C + k_3 B$$

where A, B, C is the concentration of the corresponding components.

For certainty* one may choose an initial state of the particle system setting condition

---

*To analyze (4.2) the first two equations are sufficient since concentrations A_i are connected by the conservation condition of the number of particles $C_0 = \Sigma_{i=1} A_i$.

$C(0) = C_0, \quad A(0) = B(0) = 0$

Let us discuss the formation of k_{ex} on the basis of the introduced values k_i. We shall analyze system (4.2) using the Laplace transform

$$a(s) = \int_0^\infty A(t)e^{-st}dt, \qquad b(s) = \int_0^\infty B(t)e^{-st}dt$$

$$c(s) = \int_0^\infty C(t)e^{-st}dt$$

Then, system (4.2) takes the form

$$sa(s) = -k_1a(s) + k_2b(s)$$

$$sb(s) = -(k_2 + k_3)b(s) + k_1a(s) + k_4c(s) \tag{4.3}$$

$$sc(s) = c_0 - k_4c(s) + k_3b(s)$$

From the second equation we find

$$b(s) = (s + k_2 + k_3) = k_1a(s) + k_4c(s)$$

i.e.,

$$b(s) = (k_1a + k_4c)(s + k_2 + k_3)^{-1}$$

Substituting the obtained expression for $b(s)$ into the firest equation, one derives a kinetic equation determining the flux for a direct and reverse reaction:

$$sa(s) = -k_1a(s)\frac{k_3 + s}{k_2 + k_3 + s} + k_2c(s)\frac{k_4}{k_2 + k_3 + s} \tag{4.4}$$

The equilibrium condition between initial (C) and final (A) components is readily calculated from (4.4). In limit $s \to 0$ (this limit for values $a(s)$, $b(s)$, and $c(s)$ conforms to limit $t \to \infty$ for the corresponding concentrations A, B and C) for the ratio $a(0)c^{-1}(0) = A_{eq}C_{eq}^{-1}$ (index eq points to the limit equilibrium value of concentrations at $t \to \infty$) one obtains

$$\frac{A_{eq}}{C_{eq}} = \frac{k_2k_4}{k_1k_3} = k_{eq} \tag{4.5}$$

and as it follows from (4.2)

$$\frac{A_{eq}}{B_{eq}} = \frac{k_2}{k_1}, \qquad \frac{B_{eq}}{C_{eq}} = \frac{k_4}{k_3}$$

Value K_{eq} defines the equilibrium between initial and final products of the reaction.

Let us clarify the conditions under which Equation (4.4) formally describes the diffusion kinetics of Smoluchowski with stationary rate constant k_D when concentrations of A and C particles are connected by equation

$$dA(t)/dt = k_D C(t) \tag{4.6}$$

It can be easily proved that if $c(s)$ and $a(s)$ are values of the same order and if inequalities

$$k_2 k_4 \gg k_1 k_3, \qquad s \ll k_2 + k_3 \tag{4.7}$$

at which

$$sa(s) = \frac{k_2 k_4}{k_2 + k_3} \; c(s)$$

are satisfied, then one can deduce relation (4.6) from (4.4) but with a general rate constant of the process

$$k_{ex} = k_2 k_4/(k_2 + k_3) \tag{4.8}$$

instead of k_D.

The formally introduced intermediate component B is missing in the usually studied DCR picture. It acquires a definite meaning if by B particles one understands particle pairs: mobile particle-traps which have come to a "contiguity" necessary for the subsequent chemical transformation, but have not yet experienced the latter. Preserving the previous terminology it is possible to connect B particles with a complex of interacting particles in a cage (particles in the process of collision).

In the general case it is necessary to introduce a special state in phase space (later it will be called state II), which makes the particles appearing in it appear to turn into particles of another type, but yet different from the final product. Such an interpretation of component B means that reaction $B \rightleftarrows C$ characterizes the diffusion stage of DCR.

Now let us once more turn to expression (4.8). This expression includes three constants, and the addition law of kinetic resistances does not follow from value k_{ex}. This law can be obtained given that $k_3 = k_4 = k_D$. In this case, condition (4.7) means $k_2 \gg k_1$ and it can be assumed that constant k_2 characterizes the rate of chemical transformation, i.e., $k_2 \equiv k$, while constant k_1 is the rate of the reverse chemical reaction.

If at the same time, it is demanded that $k \gg k_D$, one obtains (transferring DCR into a one-stage reaction):

$$k_{ex} = k_4 = k_D \tag{4.9}$$

in zero approximation in small parameter $\eta = k_D/k = k_3/k_2$.

Thus, the Smoluchowski method with boundary condition of the I type on the surface of the absorbing particle actually considers two successive irreversible reactions as one diffusion stage:

$$A \xleftarrow{k} B \xleftarrow{k_D} C, \text{ or more exact } A \xleftarrow[\text{boundary}]{k_D} C \quad (k_D \ll k) \tag{4.10}$$

The introduction of a boundary condition of type III as has been suggested by Collins and Kimball [29] means the explicit consideration and introduction of a reverse reaction for the diffusion stage ($A \leftarrow B \rightleftarrows C$) with the same rate constants for the direct and reverse reactions.

Note that equality $k_3 = k_4$ and the fact that rate constant k_{ex} is calculated in zero order in small parameter $\eta = k_D/k_2$, is usually not stressed in discussing the genesis of the I-type boundary condition. This hinders the adequate analysis of the reverse reaction and equilibrium conditions (4.5). The effective rate of the reverse reaction is found from Equation (4.4) provided that $k_1k_3 \gg k_2k_4$ (as before we assume that $a(s)$ and $c(s)$ are values of the same order):

$$sa(s) \approx -k_1 \frac{k_D}{k_2 + k_D} a(s) \tag{4.11}$$

Expression (4.11) defines the decomposition rate of component A (the formation of mobile particle C and an empty trap) with effective rate constant

$$k_{ex}^{rev} = \frac{k_1k_D}{k_2 + k_D} \approx \frac{k_1k_D}{k_2} \approx k_1\eta \tag{4.12}$$

Value k_{ex}^{rev} is proportional to parameter $\eta (\eta \ll 1)$ and is small as compared to k_1. Therefore, if the rate of the direct reaction is set by Equation (4.9), it is insufficient to introduce rate constant k_1 of the reverse reaction of the chemical transformation for considering the reverse process: in zero order in parameter η value $k_{ex}^{rev} = 0$. This is quite natural since it is essential "to open" the diffusion stage of the reverse reaction also in order to ensure the proceeding overall reverse reaction. As has seen already, in first order in η, one obtains $k_{ex} \approx k_D$, $k_{ex}^{rev} \approx k_1k_Dk^{-1} \approx k_1\eta$ which guarantees the satisfaction of Equation (4.5)*.

---

*However, the addition Law of Kinetic Resistances is not yet evident.

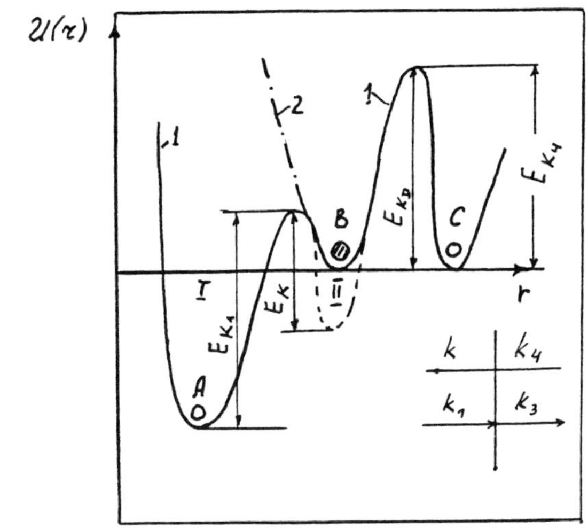

Fig. 4.1 Profile of a potential energy surface for two successive monomolecular reactions.

Thus, though the rate of the direct reaction (which yields Equation (4.9) can be calculated on the basis of the Smoluchowski method with a Type I boundary condition, calculation of the reverse reaction rate requires a more complicated model which provides for obtaining the Type III Collins-Kimball boundary condition.

A change in the notion of the DCR mechanism, including consideration of Collins-Kimball boundary conditions instead of Smoluchowski ones, can be illustrated using the representations of reaction potential surfaces [14,20,84]. For two successive monomolecular reactions, the section of potential surface along the coordinate of the reaction can be presented by the diagram shown in Figure 4.1.

Let us present the reaction path in the form of a succession of several states (I,II,C, see Figure 4.1) characterized by the potential energy minima of the considered system of reagents. Let the particle of initial component C be in state C_1, in State II—particles of component B (State II can be interpreted as the intermediate state of reaction $C \rightarrow A$), in State I—particles captured by traps (State I can be considered to correspond to the reaction product).

The course of DCR can be represented as a transition of the system from State C_1 into State I through State II. In passing from State C_1 into State II the system overcomes energy barrier E_{k_D}; from State II into State I—the barrier of the chemical transformation proper E_k. Thus transition $C_1 \rightarrow$ II characterizes the completion of the DCR diffusion stage, while transition II $\rightarrow$ I corresponds to the chemical stage.

The relation between the rates of diffusion and reaction is determined by the relative positions of potential minima, characterizing States C_1, II and I. In particular, if minimum II is located lower than minimum C_1

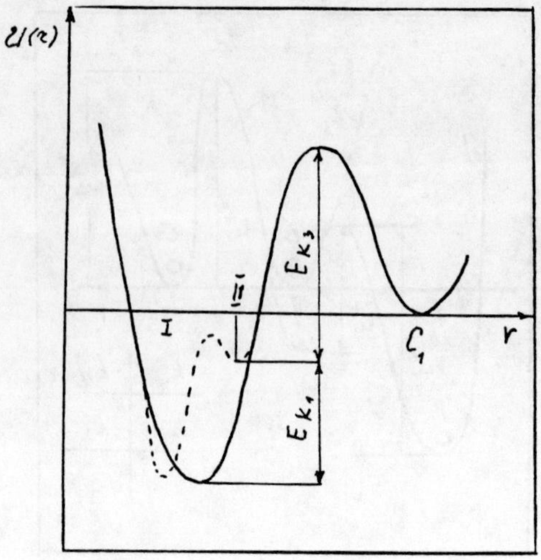

Fig. 4.2 Profile of the potential energy surface for reactions (4.10).

(dashed curve) the rate constants of the direct and effective reverse reactions of the diffusion stage are different: $k_D = k_4 \neq k_3$. And if in addition $E_k \ll E_{k_1}$ the reaction practically becomes a one-stage type. This case is illustrated in Figure 4.2 (full curve). Namely, this case corresponds to the Smoluchowski boundary condition.

The limiting stage of process $A \leftarrow C$ is defined by value E_k relative to $E_{k_D} \equiv E_{k_4}$. If $E_k \gg E_{k_4}$ (Curve 2, Figure 4.1) the limiting stage is chemical, if $E_k \ll E_{k_4}$, it is diffusional. Type III boundary conditions must correspond to the latter case (Curve I, Figure 4.1).

Now let us consider in more detail the change of representations in passing from the Smoluchowski problem to the Collins-Kimball problem.

The formulation of these problems includes an analysis of processes passing on the boundary, though the notion boundary is not always introduced consistently. Within the framwork of the analogy drawn between DCR and formal kinetics, it is logical to define the position of the boundary by State II. However, in the Smoluchowski method, when it is assumed that $k \gg k_4 = k_3 = k_D$, State II may be "lost" at a qualitative analysis. Then the reaction must proceed directly from State C_1, which is actually the boundary of the diffusion region. Namely, this fact stipulates the known difficulties of the Smoluchowski theory connected with the divergence of the flux at short times ($t \to 0$) and with the requirements following from the principle of detailed equilibrium.

4.2. MODEL REPRESENTATIONS OF DCR POTENTIAL SURFACES

The given considerations following from formal-kinetic arguments may be further substantiated by analyzing a simple diffusion controlled reaction

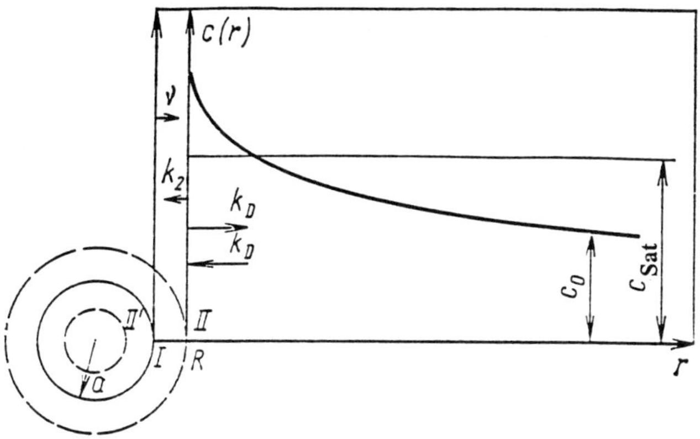

Fig. 4.3 Diagram of concentration distribution for the system particle—nonsaturated solution.

which is traditionally used [61] for approbating the theory of dissolution DCR-processes [108].

For simplicity let us consider a stationary process (at spheric symmetry of the system). Let us place into the origin of coordinates a spheric (of radius R) macroparticle of a solid component immersed into a solution of the same component. In a solution having a concentration higher than equilibrium (an over-saturated solution) the particle of the dissolved substance absorbs molecules from the solution thereby increasing its diameter. If radius R is sufficiently large so that the change in the size of the macroparticle during observation time can be ignored, the problem of the absorption rate of dissolved molecules by the particle (the growth rate of macroparticle mass) is reduced to the trap problem considered by Smoluchowski. In our case, it is more convenient to consider a non-saturated solution when the spheric particle dissolves. The observed concentration distribution in a solution is shown in Figure 4.3. In the stationary case for concentration $c(r)$ of substance molecules in a solution, the following equation is valid:

$$d^2(rc)/dr^2 = 0 \tag{4.13}$$

with the condition of a constant concentration on the surface of the macroparticle:

$$c(R) = c_a \tag{4.14}$$

and at infinity:

$$c(\infty) = c_0 \tag{4.15}$$

The solution of problem (4.13–4.15) is easily found

$$c(r) = c_0\left(1 - \frac{(c_0 - c_a)R}{rc_0}\right),$$

$$\frac{dc}{dr} = \frac{R}{r^2}(c_0 - c_a)$$

Diffusion flux $j_{\rightarrow}$ in the direction of the r increase (from the macroparticle) is equal to

$$j_{\rightarrow} = -4\pi DR^2 \left.\frac{dc}{dr}\right|_{r=R} = 4\pi DR(c_a - c_0)$$

If the flux is considered the rate characteristic of the dissolution diffusion stage, the rate constant of this stage $k_3 = k_D = 4\pi DR$ coincides with the rate constant of the direct reaction (absorption of macroparticle mass from an oversaturated solution) calculated according to the Smoluchowski method with conditions $c(R) = 0$, $c(\infty) = c_0$.

Consideration of these problems naturally gives rise to the question of how one can describe the stage of the chemical transformation as such, i.e., the precipitation—dissolution process?

Usually [61] it is done in the following way. We introduce molecule surface concentration n_s [as distinct from bulk concentration $c(r)$], probability ν of particle removal from the surface per time unit, and mean velocity u_s characterizing the return of molecules from the solution and their sticking to the sphere of radius R. Then rate j_{chem} of the dissolution-precipitation process on the surface of the sphere can be presented in the form

$$j_{chem} = 4\pi R^2(n_s\nu - c_a u_s) \tag{4.16}$$

Now one can easily derive the precipitation rate constant k_2 (the direct chemical reaction): $k_2 = 4\pi R^2 u_s$. Constant k_1 of the dissolution reaction (the reverse reaction) is usually determined in the following way [61]. Since expression $j_{chem}^{rev} = 4\pi R^2 n_s \nu$ contains surface concentrations, we introduce value c_{sat}—the equilibrium concentration of saturation:

$$n_s\nu = c_{sat}u_s, \qquad c_{sat} = c_a^{eq} \tag{4.17}$$

at which $j_{chem}^{eq} = 0$. Then

$$j_{chem} = 4\pi R^2 u_s(c_{sat} - c_a)$$

i.e.,

$$k_1 = 4\pi R^2 u_s = k_2$$

Condition (4.17) shows the absence of the concentration gradient in an equilibrium state up to the boundary. In the general case such a gradient is present, and then

$$j_{\text{dif}} = -\ 4\pi D R^2 \left.\frac{\partial c}{\partial r}\right|_{r=R} \neq k_2 c_a^{\text{eq}}$$

Let us discuss Equation (4.16) in more detail and try to combine the diffusional and formal-kinetic description of the dissolution-precipitation process. For convenience we shall try to attach meaning in terms of diffusion kinetics to intermediate product B, which is introduced in formal kinetics (4.1). As has been mentioned above, a B particle localized in the interphase boundary region can be presented as a complex of interacting particles (in our case a "evaporating" particle connected with the surface of a macroparticle) capable of further transformation, in particular, of particle injection into the diffusion region (in the case of the macroparticle dissolution problem). This actually means that there is an inner structure of the interphase boundary [200]. Let us use the representations of paper [200] and treat the interphase boundary as a combination of two states for the particles appearing on the boundary: States I and II shown in Figure 4.3. State II, characterized by volume particle density c_a, is the "boundary state" of the diffusion region*. Its specificity is determined by the fact that the particles existing in this state may quit it by two channels: from the macroparticle by the mechanism determining diffusion mobility and to the macroparticle having reacted with rate k_2. Further discussion requires complicating the notion of the surface of the absorbing particle [200].

Let us introduce (see Figure 4.3) the true radius of a macroparticle ($a < R$) determining the surface where surface concentration n_s is given and included into (4.16). The state of the particles on the surface is denoted by State I. Particles in State I can pass into State II (to the right) with probability ν, as well as with probability ν_1 pass into State II (to the left) of region $r < a$ characterized by volume particle density in a macroparticle.

---

*The conditionality of such an approach is evident. When considering models of stochastic transitions between discrete states one cannot represent the boundary of the diffusion region only by the discrete state, since in the diffusion limit a large number of discrete levels in the vicinity of the interphase boundary (at $r > R$) forms diffusing particle concentration on the surface. Nevertheless the introduction of States I and II has heuristic value. A consistent solution of the problem in the model of discrete random walks (see Section 4.3) leads to the same boundary conditions as the considerations according to [200].

Using the notion of States I, II and II', one may discern particles located in the diffusion region inside the macroparticle and in each of States I, II, II'. For coordination with the scheme of formal kinetics, it is convenient to consider particles in State I as particles of type *A* (i.e., the product), particles in State II—as *B* type particles and particles in diffusion volume—as initial component *C*. Such an identification evidently corresponds to Scheme 4.1 of the general reaction. By "collision" one can understand the formation of intermediate product *B*, which is in State II.

The main merit of treating the dissolution-precipitation process as a case of the diffusion controlled process is the possibility of discussing the macroscopic process using elements of microscopic theory. In the case of a purely microscopic consideration, the notion of an absorbing boundary loses sense, but the notion of States I, II and II' must be preserved. Then the notion of a diffusion region becomes more complicated, since instead of the notion of particle concentration one should use the probability interpretation of particle transfer dynamics. It should also be taken into account that the solvent (or the medium in which mass transfer takes place) in the microscopic approach cannot be treated as a structureless continuum since elementary jumps defining transfer displacements are thermally activated processes (as well as the chemical reaction). In this case, transfer transitions and the chemical transformation proper can be described within the framework of a single physical concept which employs the notion of activation energy (the relations between the diffusion and reaction activation energies to a great extent determine the degree of diffusion control of the overall reaction course).

Thus, one can draw the conclusion [84] that it is convenient to consider the transfer process as a number of chemical reactions and apply approaches earlier devised for analyzing the chemical reaction kinetics for its description.

Most models of chemical reaction dynamics employ representations of the theory of absolute reaction rates [14]. Let us use them for a qualitative joint description of the transfer motion and the chemical reaction. We shall consider transfer displacements as particular cases of the chemical reaction.

The main notions in the theory of absolute reaction rates are notions about potential surfaces in multi-dimensional configuration space (in the general case—phase space) and about an activated complex, which under a number of assumptions, permits evaluating the probability for direct and reverse reactions. It is these estimates of the probability of direct and reverse processes that we will need to establish the interrelation between diffusion and reaction in the problem of boundary conditions.

Generally speaking, in the description of diffusion controlled reactions, configuration space must include the coordinates of all atoms in the system [209]. Hence, it is necessary to introduce simplifying assumptions allowing separation of the diffusion and reaction. It is convenient to clarify the origin of such model assumptions (and the models of transport and reaction) using potential energy surface profiles along the reaction path, i.e., graphs of potential energy dependence *u* on reaction coordinate *r*. Examples of how such representations are used in discussing DCR theory and the problem of boundary conditions are given in papers

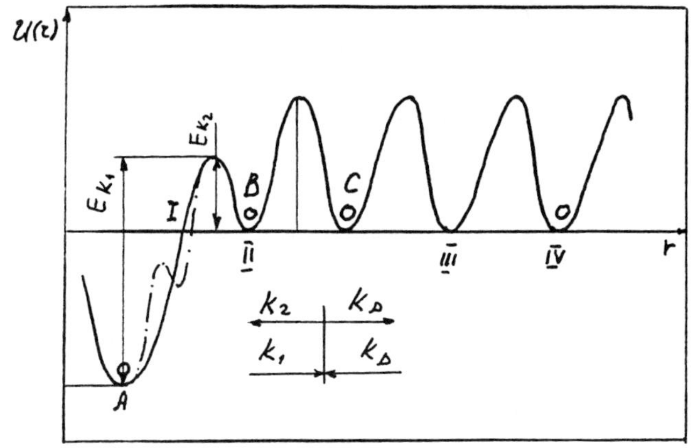

Fig. 4.4. Schematic dependence of potential energy on the reaction coordinate for migration of a mobile particle near the trap.

[206,207] where the chemical transformation kinetics in liquid media is considered.

In the case of a diffusion limited reaction, activation energy of the chemical reaction must be lower than diffusion activation energy (with the exception of cases when the principal role is played by entropy, non-activation factors). When analyzing DCR using representations of formal kinetics, see Scheme (4.1), it is convenient to choose a potential profile in the way shown in Figure 4.4, if the diffusion control of the process is realized. Such diagrams are most frequently found in discussions of reactions and diffusion in the solid phase. In particular, a diagram (Figure 4.4) may illustrate the relations between energy $F = E_{k_1} - E_{k_2}$ of defect formation and diffusion activation energy E_D. More general diagrams, this time for DCR of point defects, are given in Figure 4.5 and Figure 4.6. Figure 4.5 is used in paper [162] for analyzing the mobility and reactions of inter-site atoms and vacancies on the basis of classical and diffusion kinetics. The region including minima u_1, u_2, A, B characterizes the distortion of the crystal lattice near the vacancy and clarifies the possibility of introducing the type I states. The potential diagram in Figure 4.6 has been used [210] for analyzing the defect annealing kinetics in the presence of an absorbing surface of a spheric sink when there is a barrier between the bulk material and the sink. The nature of this barrier for defect migration to a sink can be determined by local stress fields or the presence of impurities near the absorbing sink surface.

The diagram in Figure 4.4 will be considered as the reference in designing DCR models by methods using random process theory. It permits attaching a certain sense to the notions of collision, boundary of the absorbing medium or particle, transition layer width, etc. Reflecting the possible relations between reaction and diffusion in many-phase media, the diagram permits taking into account reversibility effects and the detailed balance principle. When applying this diagram in the random walk method, potential minima are presented by points of a lattice whose boundary is

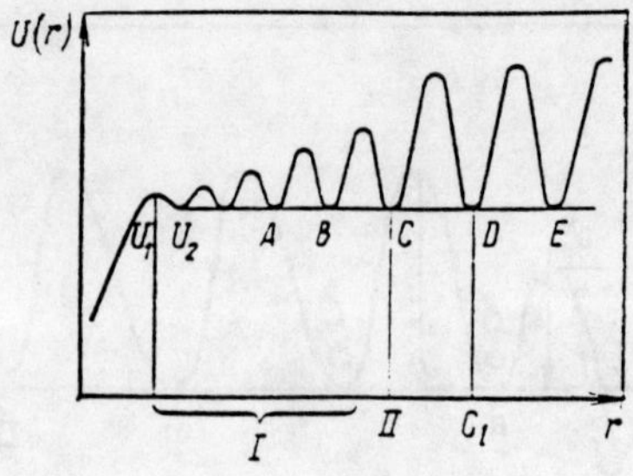

Fig. 4.5. Dependence of potential energy on distance for migration of an intersite atom near a vacancy [162].

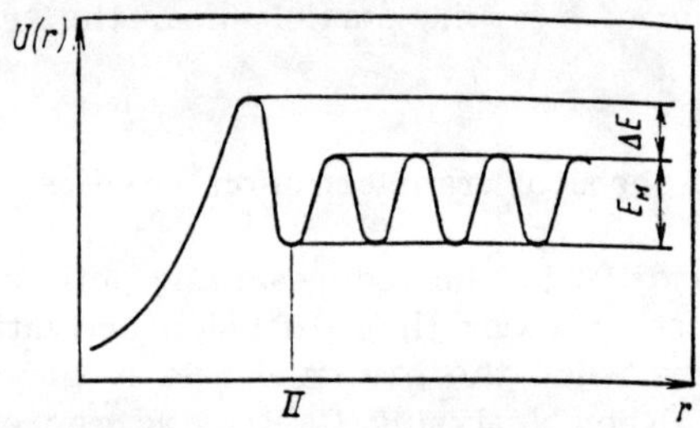

Fig. 4.6. Dependence of potential energy on distance for the case when activation energy E_m of the defect migration increases by value ΔE for the particle jump to the absorbing sink surface [210].

defined by States I and II. The derivation of boundary conditions obtained from considering discrete processes of random walks in such a lattice should yield general relations between probabilities for diffusion and reaction, while the system of representations corresponding to Figure 4.4 allows expressing these probabilities through the molecular parameters of the system.

4.3. GENERAL FORM OF BOUNDARY CONDITIONS [211]

We shall start deriving boundary conditions of a general form by representing the boundary as the combination of States I and II described in Section 4.1.

Let us consider the discrete process of random walks. We shall assume that a particle moves randomly through the points of a one-dimensional lattice so that during short time τ_0 of jump it passes from discrete state j reached by time l (l is the full number of jumps) into neighboring states $j \pm 1$. Let us denote by $n_{j,l}$ the probability of finding the particle in

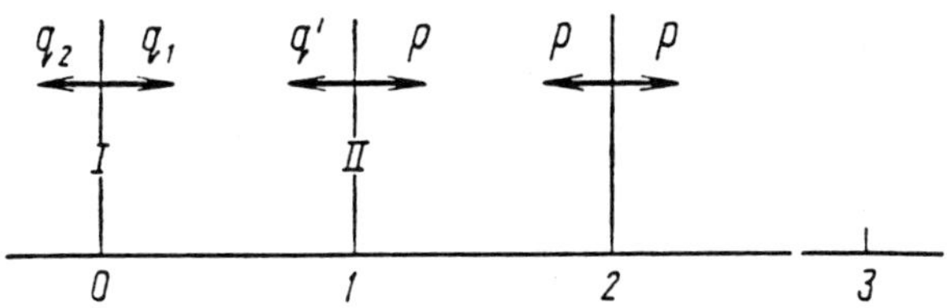

Fig. 4.7. Scheme of transition probabilities for random walk processes.

point j after l jumps. Then the process can be described by the following system of equations:

$$n_{j,\,l+1} = p(n_{j-1,l} + n_{j+1,l}), \qquad j > 1 \tag{4.18}$$

$$n_{1,l+1} = q_1 n_{0,l} + p n_{2,l} + (1 - p - q')n_{1,l} \tag{4.19}$$

$$n_{0,l+1} = (1 - q_1 - q_2)n_{0,l} + n_{1,l}q' \tag{4.20}$$

where p is the probability of transitions $j \rightarrow j \pm 1$ (at $j \geqslant 2$); q_1 and q' are the probabilities of transitions $0 \rightarrow 1$, and $1 \rightarrow 0$; q_2 is the probability for leaving the lattice from state 0.

The scheme of transition probabilities is given in Figure 4.7. The essence of the scheme is as follows. According to Section 4.1, the diffusion region is limited by State II, which corresponds to the first point of the lattice $j = 1$. State I, characterizing the reaction product, corresponds to lattice point $j = 0$. For simplicity, probability p of the transition from point to point in the diffusion region is assumed to be constant $p = 1/2$. The relation between transition probabilities between points $j = 0$ and $j = 1$ determines various physical stochastic processes. In particular, the case $q_2 = 0$ may serve as a model for describing a reversible monomolecular DCR; the case $q_1 = q' = p$, $q_2 \neq 0$ is a model for a monomolecular irreversible reaction, when the probabilities for the diffusional jump and the chemical transformation proper coincide.

Let us first consider various approximate equations obtained for expansion to a series of Equations (4.18) of system evolution. Introducing continuous variables

$$x = j\varepsilon, \qquad t = l\tau_0, \qquad n_{j,l} = \varepsilon\rho(x, t) \tag{4.21}$$

where ε is the size of the jump, again assumed to be small as compared to the characteristic parameters of the system, one obtains

$$\rho(x, t + \tau) = p[\rho(x - \varepsilon, t) + \rho(x + \varepsilon, t)]$$

instead of Equation (4.18).

Expanding density ρ in its Taylor series we derive

$$\rho(x, t) + \frac{\partial \rho(x, t)}{\partial t} \tau_0 + \frac{1}{2} \frac{\partial^2 \rho(x, t)}{\partial t^2} =$$

$$= \frac{1}{2} \left[2\rho(x, t) + \frac{\partial^2 \rho(x, t)}{\partial x^2} \varepsilon^2 \right]$$

Preserving terms of first order in τ_0 one obtains the diffusion equation

$$\frac{\partial \rho(x, t)}{\partial t} = \frac{\varepsilon^2}{2\tau_0} \frac{\partial^2 \rho(x, t)}{\partial x^2} = D \frac{\partial^2 \rho(x, t)}{\partial x^2} \tag{4.22}$$

if the diffusion coefficient is defined by the usual relation

$$D = \frac{\varepsilon^2}{2\tau_0} \tag{4.23}$$

In the second order in τ_0 one obtains the telegrapher's equation [66,212] (1.110)

$$\frac{\partial^2 \rho}{\partial t^2} + \frac{2}{\tau_0} \frac{\partial \rho}{\partial t} = \frac{\varepsilon^2}{\tau_0^2} \frac{\partial^2 \rho}{\partial x^2} \tag{4.24}$$

assuming

$$\varepsilon^2/\tau_0^2 = \nu^2, \qquad 2/\tau_0 = \beta_0 \tag{4.25}$$

Value β_0 is usually defined as the friction coefficient, while coefficient ν^2 characterizes the displacement "velocity" during elementary time and it is connected with coefficients of diffusion D and friction β_0 by the relation

$$\nu^2 = \varepsilon^2/\tau_0^2 = D\beta_0$$

It is seen that Equation (4.24) is far more general than the usual diffusion Equation (4.22). If friction coefficient β_0 is small, $\beta_0 \to 0$ Equation (4.24) is reduced to a wave equation describing the spatial distribution of particles free from any interactions with the medium or with each other.

The preservation of the first term from the left (i.e., the second time derivative) at $\beta_0 \neq 0$ may be of principal importance [212] (see Section 1.8). For example when solving (4.24) with discontinuous initial conditions, no problems on the "infinite" diffusion rate arise.

Using Equation (4.18) it is also possible to describe semicontinuous random processes when the discreteness of the process over the spatial variable is essential. In particular, it is important for studying the succession of chemical reactions in formal kinetics.

Passing from (4.18–4.20) to continuous time variable $t = l\tau_0$ one derives

$$\frac{\partial N_j(t)}{\partial t} = k_D[N_{j-1}(t) + N_{j+1}(t) - 2N_j(t)] \tag{4.26}$$

where $k_D = (2\tau_0)^{-1}$ is the probability of a transition from point to point per time unit in the diffusion region.

The boundary conditions assume the form

$$\frac{\partial N_1}{\partial t} = k_1 N_0(t) + k_D N_2(t) - (k_D + k_2)N_1(t) \tag{4.27}$$

$$\frac{\partial N_0}{\partial t} = k_2 N_1(t) - (k_1 + k_4)N_0(t) \tag{4.28}$$

where

$$k_1 = q_1\tau_0^{-1}, \qquad k_4 = q_2\tau_0^{-1}, \qquad k_2 = q'\tau_0^{-1} \tag{4.29}$$

Thus the system of Equations (4.18–4.20) possesses sufficient generality for determining the type of boundary conditions. Let us return to an analysis of this general system. We shall first find its formal solution and elucidate the asymptotic behavior of this solution at large j and l (at $j \ll l$). Proceeding to continuous variables in the asymptotic solution is quite correct and we obtain the continuous analogue of the exact solution. After that we shall analyze the relation between the derivative and the function in the vicinity of the boundary.

The system (4.18–4.20) will be solved with initial conditions

$$n_{j,0} = b = \text{const}(n_{j,l} \longrightarrow b \text{ at } j \longrightarrow \infty) \tag{4.30}$$

$$n_{0,0} = 0$$

Condition $n_{0,0} = 0$ points to the fact that at the initial time moment State I contains no particles.

To solve the system, we introduce generating function

$$n_j(s) = \sum_{l=0}^{\infty} n_{j,l}s^l \tag{4.31}$$

for which follows from (4.18) and (4.30) equation

$$n_j(s) - b = ps[n_{j-1}(s) + n_{j+1}(s)] \tag{4.32}$$

and boundary conditions

$$n_1(s) - b = n_0(s)q_1 s + psn_2(s) + s(p - q')n_1(s)$$

$$n_0(s) = s(1 - q_1 - q_2)n_0(s) + q'sn_1(s)$$

Excluding $n_0(s)$ from the last pair of equations we obtain

$$n_0(s) = q'sn_1(s)/(1 - \bar{q}s) \tag{4.33}$$

$$n_1(s)\left[1 - q_1 s\frac{q's}{1 - s\bar{q}} - s(p - q')\right] = b + spn_2(s) \tag{4.34}$$

where $q = q_1 + q_2$, $\bar{q} = 1 - q$.

Now Equation (4.34) plays the role of a boundary condition for the random walk process described by Equation (4.32).

The general solution of system (4.32) is easily found [163,165]; it has the form

$$n_j(s) = A + \beta^j B, \qquad A = b(1 - s)^{-1}$$

where $\beta = s^{-1}(1 - \sqrt{1 - s^2}) \leqslant 1$; constant B is found from condition (4.34).

And the final result is

$$n_j(s) = \frac{l}{1 - s}\left\{1 - \beta^j\frac{q's(1 - s + sq_2)}{sp(1 - s\bar{q}) - \beta s[(p - q')(1 - sq) + q_1 q's]}\right\} \tag{4.35}$$

To proceed to macroscopic variables, one has to find the form of $n_{j,l}$ at large j and l which corresponds to searching for the asymptotic behavior of $n_j(s)$ at $s \to 1$.

Let $s = e^{-\delta}$ at $\delta \ll 1$. Then $\beta = 1 - (2\delta)^{1/2}$, $\beta^j = \exp[-j(2\delta)^{1/2}]$, and proceeding to the continuous variable in (4.31) one obtains from (4.35) and (4.33):

$$n_j(\delta) = \frac{b}{\delta}\left[1 - \frac{q'(\delta + q_2)\exp(-j\sqrt{2\delta})}{q'(\delta + q_2) + p\sqrt{2\delta}(q_1 + q_2 + \delta)}\right] \tag{4.36}$$

$$n_0(s) = \frac{b}{\delta}\frac{q'}{(\delta + q_1 + q_2)}\left[1 - \frac{q'(\delta + q_2)(1 - \sqrt{2\delta})}{q'(\delta + q_2) + p\sqrt{2\delta}(q_1 + q_2 + \delta)}\right] \tag{4.37}$$

So in (4.31) it is possible (at $l \gg 1$) to approximate the sum by an integral, i.e., to treat (4.31) as a Laplace transform for $n_{j,l}$. At $j \gg 1$ variable j may be considered continuous. Then derivative $\partial n_j(\delta)\partial_j$ is readily calculated from (4.36).

Now let us clarify how the asymptotic form of $n_j(\delta)$ approximates the situation in the vicinity of the boundary. One obtains from (4.36)

$$N(0,\delta) = n_j(\delta)\Big|_{j=0} = \frac{b}{\delta}\, p\sqrt{2\delta}\; \frac{\delta + q_1 + q_2}{q'(\delta + q_2) + p\sqrt{2\delta}\,(\delta + q_1 + q_2)} \tag{4.38}$$

$$\frac{\partial N(0,\delta)}{\partial_j} = \frac{\partial n_j(\delta)}{\partial_j}\Bigg|_{j=0} =$$

$$= \frac{b}{\delta}\sqrt{2\delta}\; \frac{q'(\delta + q_2)}{q'(\delta + q_2) + p\sqrt{2\delta}\,(s + q_1 + q_2)}$$

($j = 0$ means $j = 0(1)$).

The exact expression for the population of the zero level gives expression (4.37); it follows from comparing (4.37) and (4.38) that

$$n_0(\delta) = \frac{q'}{\delta + q_1 + q_2}\left[N(0,\delta) + \frac{\partial N(0,\delta)}{\partial j}\right] \tag{4.39}$$

It also follows from (4.38) that

$$q'(\delta + q_2)N(0,\delta) = \frac{\partial N(0,\delta)}{\partial j}\, p(\delta + q_1 + q_2) \tag{4.40}$$

If we regard δ as a variable of the Laplace transform, then

$$\frac{\partial n_0(l)}{\partial l} = -(q_1 + q_2)n_0(l) + q'N(0,l) + q'\,\frac{\partial N(0,l)}{\partial j} \tag{4.41}$$

Proceeding to continuous variables $t = l\tau_0$, $x = j\varepsilon$ and using notations (4.23) and (4.29), one can derive from (4.41) an equation for the population of State I:

$$dn_0(t)/dt = -(k_1 + k_4)n_0(t) + c\rho(\varepsilon,t) \tag{4.42}$$

where $c = k_2\varepsilon$.

It should be stressed that densities $n_0(t)$ and $\rho(\varepsilon,t)$ have a different dimensionality; in the general case $n_0(t)$ is surface density, while $\rho(\varepsilon,t)$ is bulk density.

Similar transformations in Equation (4.40) defining the population of State II gives

$$D \frac{\partial \rho(\varepsilon,t)}{\partial x} = -k_1 n_0(t) + c\rho(\varepsilon,t) \tag{4.43}$$

The system of Equations (4.42) and (4.43) represent the general boundary condition for transfer equations if the latter ones are obtained using diffusion approximations. In particular, it implies a non-stationary boundary condition [201] and a Type III boundary condition. And, indeed, substituting (4.43) into (4.42) one derives equation

$$\frac{dn_0}{dt} = -k_4 n_0 + D\partial \rho(\varepsilon,t)/\partial x \tag{4.44}$$

In the limit $k_1 \to \infty$ and $c \to \infty$ it follows from (4.43) that

$$k_1 n_0 = c\rho(\varepsilon,t)$$

which upon sustitution into (4.44) leads to the non-stationary boundary condition

$$ck_1^{-1} \frac{\partial \rho(\varepsilon,t)}{\partial t} = -ck_4 k_1^{-1} \rho(\varepsilon,t) + D \frac{\partial \rho(\varepsilon,t)}{\partial x}$$

Limit $ck_1 \to 0$, $ck_4 k_1^{-1}$—finite corresponds to Type III boundary condition i.e.,

$$ck_4 k_1^{-1} \rho(\varepsilon,t) = D \frac{\partial \rho(\varepsilon,t)}{\partial x} \tag{4.45}$$

The Smoluchowski boundary condition $\rho(\varepsilon,t) = 0$ follows from (4.45) using the limit

$$k_1 D/ck_4 \longrightarrow 0 \tag{4.46}$$

We see that this condition is adequate to a changed equilibrium of the system, when rate ck_4 of the direct DCR is much higher than the rate of the reverse reaction Dk_1. This becomes especially clear if we use the analogy of DCR and formal-kinetic scheme (4.2) of successive reactions when condition (4.46) is analogous to (4.7).

4.4. SMOLUCHOWSKI PROBLEM WITH A GENERAL-TYPE BOUNDARY CONDITION

To illustrate the influence of boundary conditions upon DCR kinetics, let us consider the three-dimensional problem:

$$\partial (r\rho)/\partial t = \partial^2 (r\rho)/\partial r^2 \tag{4.47}$$

with initial condition $\rho(r,0) = \rho_0$, $n_0(0) = 0$ and boundary conditions

$$dn_0/dt = -\lambda n_0 + c\rho(R, t), \qquad \lambda = k_1 + k_4$$

$$k_1 n_0 + D\,\frac{\partial \rho(R,t)}{\partial r} = c\rho(R,t) \tag{4.48}$$

where R is the radius of particles.

The effective rate constant k_{ex} of the reaction will be introduced in the following way

$$k_{ex}(t) = k_4 n_0 \rho_0^{-1}\, 4\pi R^2 \tag{4.49}$$

The thus determined constant $k_{ex}(t)$ within the stationary limit (the case of large times) has the sense of rate constant $\bar{k}$ appearing in the Collins-Kimball Type III boundary condition (1.19)

$$k_{ex}(\infty) = \bar{k} = 4\pi R^2 k_4 c \lambda^{-1} \tag{4.50}$$

Here we have taken into account the fact that according to (4.50) $n_0/\rho(R,\infty) = c/\lambda$. Relation (4.50) reveals the microscopic sense of constant $\bar{k}$.

We shall now show that value $k_{ex}(t)$ does not diverge at $t \to 0$ (unlike the rate constant in the Smoluchowski method). For solving the non-stationary problem we introduce function $W(r,t) = r\rho(r,t)$, new variable $x = r - R$ and use the Laplace transform

$$W(x,\delta) = \int_0^\infty W(x,t)\,e^{-\delta t}\,dt \tag{4.51}$$

Then instead of (4.47) and (4.48) we obtain

$$\delta W(x,\delta) = D\,\frac{\partial^2 W}{\partial x^2} + \rho_0(x + R)$$

$$\delta n_0(\delta) = -\lambda n_0(\delta) + cR^{-1}W(0,\delta) \tag{4.52}$$

$$k_1 n_0(\delta) = cR^{-1}W(0,\delta) - DR^{-1}\,\frac{\partial W(0,\delta)}{\partial x} - R^{-1}W(0,\delta) \tag{4.53}$$

The solution of Equation (4.52) has the form

$$W(x,\delta) = B\exp\left(-x\sqrt{\delta/D}\right) + \rho_0\delta^{-1}(x + R) \tag{4.54}$$

Here B is the constant determined from boundary conditions (4.53):

$$B = -\frac{\rho_0 cR^2(\delta + k_4)}{\delta\left[Re(\delta + k_4) + (\delta + \lambda)D\left(1 + \sqrt{R^2\delta/D}\right)\right]} \tag{4.55}$$

Within the limit of short times at $\delta \gg \lambda$, one derives from (4.55)

$$B = -\rho_0 cR^2 \left[cR\delta + \delta D\left(1 + \sqrt{R^2\delta/D}\right)\right]^{-1}$$

Let us simplify this expression assuming the reaction is diffusion limited, i.e., $cR/D \gg 1$. Then

$$B \approx -\rho_0 cR^2\delta^{-1}\left(cR + D\sqrt{R^2\delta/D}\right)^{-1}$$

$$n_0(\delta) = c\rho_0\delta^{-2}\left[1 - cR^2\left(cR + \sqrt{\delta DR^2}\right)^{-1}\right] \quad c\rho_0/\delta^2$$

Proceeding to the original using a reverse Laplace transformation gives $h_0(t) \sim c\rho_0 t$.

Since according to (4.48) within this limit $\rho(R,t) \approx \rho_0 + \lambda n_0 c^{-1}$ the sought after constant k_{ex} (at $t \ll \lambda^{-1} \ll R^2/D$) takes the form

$$k_{ex} \approx 4\pi R^2 k_4 ct$$

(in the Smoluchowski method $k_{ex} \sim t^{-1/2}$).

At larger times (but at $t \ll R^2/D$) the time dependences are stronger [201].

Thus investigating reaction kinetics (e.g., radical recombination) within a wide time interval and at different initial distributions, one can experimentally determine the microscopic parameters appearing in conditions (4.48) and hence represent the mechanism of the corresponding process in more detail. In particular, boundary conditions (4.48) permit taking into account the influence of reaction reversibility upon kinetic laws.

Thus, one can draw the conclusion [75] that the duration of the reaction and its reversibility influence the character of kinetic laws, including the time dependence of the effective rate constant and its value. The generalized boundary conditions (4.48) and (4.53) allow taking into account the influence of these factors. Moreover, within the framework of these conditions one can consider both homogenous and heterogenous processes and correctly take into account the mass transfer through the boundaries of the phases.

4.5. BOUNDARY CONDITIONS FOR MASS TRANSFER THROUGH THE INTERFACE

We shall confine ourselves to considering a one-dimensional problem when the phase boundary is a plane (the generalization in the two- and three-dimensional case is connected with the use of the corresponding expressions for diffusion fluxes). Upon obtaining conditions at the interface gas (region $x < a$)—condensed phase (region $x > a$) one should introduce flux I_0 hitting interface $x = a$ from region $x < a$ and determined by gas pressure. The sought after boundary conditions then have the form

$$\frac{dn_0}{dt} = -n_0\mu + I_0 + D\left.\frac{\partial\rho}{\partial x}\right|_{x=a} \tag{4.56}$$

$$\lambda n_0 = c\rho(a,t) - D\left.\frac{\partial\rho}{\partial x}\right|_{x=a} \tag{4.57}$$

Analogous equations for the high degree of surface occupation are suggested in [204].

When considering mass transfer through surface $x = a$ separating condensed media, the boundary condition is written in the form of a three equation system:

$$\frac{dn_0}{dt} = D_1\left.\frac{\partial\rho}{\partial x}\right|_{x=a+0} - D_2\left.\frac{\partial\rho}{\partial x}\right|_{x=a-0} \tag{4.58}$$

$$\lambda_1 n_0 = c_1\rho_1(a+0,t) - D_1\left.\frac{\partial\rho}{\partial x}\right|_{x=a+0} \tag{4.59}$$

$$\lambda_2 n_0 = c_2\rho_2(a-0,t) + D_2\left.\frac{\partial\rho_2}{\partial x}\right|_{x=a-0} \tag{4.60}$$

where lower indexes (1; 2) refer to regions $x > a$ and $x < a$ respectively.

From Equations (4.58–4.60) one can easily derive the microscopic representation for parameter γ usually used in the analysis of sorption dynamics when the relation $\rho_1(a+0) = \gamma\rho_2(a-0)$ is assumed. In this case $\gamma = \lambda_1 c_2/(\lambda_2 c_1)$ which is achieved at $D\lambda_{1,2} \to 0$.

One can derive conditions (4.58–4.60) considering random walk models in condensed phases (see Section 4.3) or equations of material balance during particle transfer through boundary $x = a$ (assigned by State I and two Type II states at $x = a \pm 0$ to the right and left of the boundary) taking into account the corresponding macroscopic fluxes into Type II states.

Boundary conditions (4.56) and (4.57) are the generalized conditions applied in [213] for analyzing the phenomenon of metal membrane superpenetrability, and they can be reduced to that from [213] in the limit $\lambda_{1,2} \to \infty$, values $\lambda_2/\lambda_1 \equiv \nu$ and $\lambda_2 c\lambda_1^{-1} \equiv x$ are finite.

Boundary conditions (4.56–4.60) may be used for analyzing the kinetics of sorption (cases of polymer oxidation, destruction of polymer coatings, etc.).

CHAPTER 5

PROCESSES OF MEMBRANE SEPARATION OF LIQUID AND GAS MEDIA

Many technological processes require separating liquid and gas mixtures. Neutral or charged components appearing in such mixtures differ by diffusion coefficients and equilibrium parameters of solubility in various materials. And thus, in principle, arises the possibility of effectively separating needed components in the process of diffusion transfer (diffusional drift) of mixtures through thin partitions (membranes) constructed from sufficiently penetrable materials (polymeric, inorganic). This separation principle is the basis of many technological processes of the chemical, petrochemical, gas, food, pharmaceutical, electronic and other branches of the national economy and industry.

The membrane separation of liquid and gas mixtures has been considered in a number of monographs [214–224] presenting much experimental data and various approaches to describing transfer processes of ions and molecules in different types of membranes (ion-exchanging, gas-permeable, inverse-osmatic, ultrafilter). As a rule, all these monographs describe the transfer process phenomenologically, by introducing the corresponding transfer coefficients (diffusion coefficients, penetrability constants) using assumptions of linear relations between the particle fluxes passing through the membrane and the active forces (trans-membrane gradients of particle concentration, electric potential, pressure). The latter is valid only for describing transfer processes at relatively small particle fluxes through membranes and at a "rapid" establishement of equilibrium distribution of transferred particles between the outer medium and the membrane in the interphase boundary region.

The further development of membrane technology is connected with realizing intensive regimes of mass- and electromass transfer through membranes. These processes are essential for creating industrial membrane technology the electrolytic production of caustic soda and chlorine, hydrogen recovery in ammonia and methanol production, and the effective use of membranes in organic synthesis and coal gassification. In particular, electric current densities during electrolysis of sodium chloride yielding alkali and chlorine exceed by an order current densities during electrodialysis demineralization and sink concentration. Also noteworthy is the fact that various experiments and theoretical calculations [212] show the advantages of using membrane gas-separation processes at high (over 10 MPa) and superhigh (over 100 MPa) pressures. Taking these advantages into account, it becomes necessary to devise adequate descriptions of membrane transport processes with strongly nonequilibrium and intensive regimes. In particular, when it is necessary to exceed the limits of the commonly used [214–216] thermodynamics of irreversible processes and to develop corresponding microscopic models for transmembrane transfer and transfer in the interphase boundary region [225–230].

This chapter presents several new tendencies in the development of membrane process physics and chemistry.

5.1. TRANSFER IN ION-EXCHANGE MEMBRANES

The kinetics of ion transfer through ion-exchange membranes are determined to a great extent by the distribution of ionogenic groups in the polymer matrix, as well as by the specificity of interphase boundaries in which submicrovicinity ($\sim$1 nm) causes a dramatic change in the electric potential (dipole in equilibrium conditions "Donnane" jump). In modelling transmembrane ion transfer, special attention is usually paid to the description of ion transport processes in membrane volume [215,216]. It is accepted that interphase boundaries of polymer ion-exchange membranes are characterized by low surface resistance for transferred ions. However, the specificity of the transfer of various ions through the separation boundary electrolyte solution—ion exchange membrane may partially determine membrane selectivity to ions of different types. In particular, this is true in the case of absorption of surface-active substance molecules (surfactants) on the ion-exchange membrane surface in proceeding to intensive masstransfer regimes under conditions of strongly nonstationary processes. Therefore, deeper investigation of ion transfer kinetics through interphase boundaries is doubtlessly of great interest.

Processes connected with ion transfer through boundary regions are usually taken into account in the case of a high rate of such a transfer, when at interphase boundaries equilibrium exists among the particles approaching the boundaries (then the rate of transmembrane transfer is determined by the diffusion and drift in membrane volume) [215,216,231]. They are also taken into account purely phenomenologically without revealing the physical sense of the introduced parameters, i.e., without establishing their relation to the rates of elementary processes (e.g., using the Dömkoler number $\tilde{\gamma} = \tau_D/\tau_R$ [232], where τ_D and τ_R are, respectively, the characteristic times of the diffusion and chemical reaction).

Evidently the assumption about the equilibrium character of surface transformations with transmembrane particle transport may appear untenable when considering transfer in subthin membranes: in particular, in the selective layer of the so-called asymmetric membranes, when the relative role of surface effects increases, and when considering sufficiently intensive mass transfer regimes (extraction regimes of some ions from the electrolyte, H^+ and OH^- ion generation in bipolar ion-exchange membranes, etc.). Taking into account surface processes purely phenomenologically may also appear insufficient because of the possible inadequacy of the usually-employed boundary conditions (even in the simplest problems of mass transfer without a chemical transformation on the surface it sometimes becomes necessary to use more complex boundary conditions than the Type III ones (see Section 4.3).

The description of mass transfer processes through interphase boundaries, including ion transfer, should take into account two essential factors: I). Structural nonuniformities of the phase conjugation region (dynamic characteristics of atom groups and polymer chain fragments in this region with characteristic size ∿1 nm may noticeably differ from volume values); 2). The presence of strong (in the general case random) electric fields in the region of interphase boundaries created by various charged and polar groups (including adsorbed atoms and molecules forming donor-acceptor complexes, various clusters and subparticles of other phases, etc.) [233]. It will be shown below that when considering problems of ion transfer through the boundary electrolyte-ion-exchange membrane one can design the model taking into account the above-mentioned factors introducing certain parameter dependences (rate constants of elementary processes) appearing in the boundary conditions for transfer problems. This enables considering the kinetics of ion transport through ion-exchange membranes under strongly nonequilibrium conditions.

5.1.1. Formulation of Boundary Conditions

The boundary conditions for solving mass transfer problems (we shall consider ion transfer) are (see Chapter 2) balance equations for fluxes between the states of adatoms (admolecules) on the membrane surface (State I) and boundary states (State II) in the diffusion-drift region [8,205]. Let us first discuss the simplest case—the transfer of ion K through the boundary separating the media. Let plane-parallel membrane occupy space region $0 < x < d$ (x is the coordinate). And let $I_K{}^L$ and $I_K{}^R$ denote fluxes of K ions hitting the membrane from region $x < 0$ (region L) and $x > d$ (region R) respectively. Since we shall not analyze processes of ion transfer outside the ion-exchange membrane (see for example [234]), we determine fluxes $I_K{}^{L(R)}$ through boundary concentrations $a_K{}^{L(R)}$ (activities) of ions outside the membrane (states of $II_{L(R)}{}^{(e)}$ types of L and R-regions of the electrolyte) assuming $I_K{}^{L(R)} = a_K{}^{L(R)}\alpha_K{}^{L(R)}$ (where $\alpha_K{}^{L(R)}$ are the rates of ion transition from $II_{L(R)}{}^{(e)}$ electrolyte states into surface states $I_{L(R)}$ at the boundaries of $L(R)$ electrolyte regions with the membrane). The rate of ion transition from state $I_{L(R)}$ into boundary states $II_{L(R)}{}^{(e)}$ outside the membrane and into boundary state $II_{L(R)}$ in the membrane and the reverse transition from state $II_{L(R)}$ in the membrane into state $I_{L(R)}$ will be denoted respectively by $\mu_K{}^{L(R)}$, $\lambda_K{}^{L(R)}$ and $c_K{}^{L(R)}$. Then the boundary condition on the boundary of electrolyte region L and the membrane takes the form

$$\frac{dn_K{}^L}{dt} = \alpha_K{}^L a_K{}^L - (\lambda_K{}^L + \mu_K{}^L)n_K{}^L + c_K{}^L \rho_K(+0,t)$$

$$\lambda_K^L n_K^L = c_K^L \rho_K(+0,t) + j_K(+0,t) \tag{5.1}$$

where $\rho_K(x, t)$ and $j_K(x, t)$ are respectively the concentration and the flux of K ions in state I_L; t is time.

Rates $\lambda_K^{L(R)}$ and $c_K^{L(R)}$ of ion transitions depend on electric potential jumps $\varphi^{L(R)}$ in the subsurface region of the membrane (region of a spatial charge), while rates $\mu_K^{L(R)}$ and $\alpha_K^{L(R)}$ depend on the corresponding jumps of the potential in the subsurface region of the electrolyte. The latter dependencies may be taken into account within the framework of the ψ' potential theory [234]; we shall not consider these dependecies (ignoring them is justified for relatively concentrated electrolytes). To determine the dependences of $\lambda_K^{L(R)}$ and $c_K^{L(R)}$ on $\varphi^{L(R)}$ we assume that states $I_{L(R)}$ and $II_{L(R)}$ in the membrane are spatially separated by a potential barrier which is the superposition of the initial potential barrier and the dipole jump $\varphi^{L(R)}$. Let φ^L and $\bar{\varphi}^L$ be the electric potentials of the surface in states I_L and II_L; $\psi_D{}^L \equiv (e/kT)(\varphi^L - \bar{\varphi}^L)$ the given dipole jump of the electric potential in the region of ion-exchange membrane spatial charge (e is the absolute value of the electron charge; with equilibrium conditions $\psi_D{}^L \equiv \psi_{D,0}{}^L$ is the so-called Donnane potential). Then

$$\lambda_K^{L(R)} \approx \lambda_{K,0}^{L(R)} \exp\left(z_K^{L(R)} \xi_K^{L(R)} \psi_D^{L(R)}\right) \tag{5.2}$$

$$c_K^{L(R)} \approx c_{K,0}^{L(R)} \exp\left[-z_K^{L(R)}\left(1 - \xi_K^{L(R)}\right)\psi_D^{L(R)}\right] \tag{5.3}$$

where $z_K^{L(R)}$ is the ion charge (in units e); $\xi_K^{L(R)}$ is the transfer coefficient equal to the relative part of $\psi_D^{L(R)}$ corresponding to the region from the interphase boundary to the transition region (extremum of the initial potential barrier).

Note that in this method of introducing rates of interphase ion transitions, the parameters $\mu_K^{L(R)}$, $\alpha_K^{L(R)}$, $\lambda_{K,0}^{L(R)}$ and $c_{K,0}^{L(R)}$ appear to be interdependent under the conditions of equilibrium:

$$\mu_K^{L(R)} = \alpha_K^{L(R)} \lambda_{K,0}^{L(R)} / c_{K,0}^{L(R)} \sigma_{K,0} \tag{5.4}$$

Here $\sigma_{K,0}$ is the equilibrium parameter of ion K solubility (the Henry constant) in membrane volume and it is defined by relation

$$\rho_K[0(d)] = \sigma_K a_K^{L(R)} = \sigma_{K,0} \exp\left[z_K^{L(R)} \psi_{D,0}^{L(R)}\right] a_K^{L(R)} \tag{5.5}$$

Now let us consider a more complicated case, when at interphase boundaries ($x = 0$ and $x = d$) chemical transformations occur proceeding with the participation of the approaching of the membrane ions, for example, the complex-formation process. Let ion K^+ approaching the membrane from the side of the solution ($x < 0$) form complex (KB) with ionofore—complex-formator B approaching the interphase boundary from the side of the membrane ($x > 0$); and let the process of complex decomposition occur at the opposite side of the membrane ($x = d$). The balance equation (boundary conditions) should be written for surface concentrations $n_{a(b)}^{L(R)}$ of all particles on the membrane surface. If we neglect Langmuire nonlinearity stipulated by the existence of limit particle concentration on the surface and in the volume, we obtain

$$dn_a^L/dt = \alpha_{Kb}^L a_K^L n_b^L - \mu_a^L n_a^L - \lambda_a^L n_a^L + c_a^L \rho_a(+0,t) \tag{5.6}$$

$$\lambda_a^L n_a^L = c_a^L \rho_a(+0,t) + j_a(+0,t)$$

$$dn_b^L/dt = \mu_a^L n_a^L - \alpha_{Kb}^L a_K^L n_b^L + c_b^L \rho_b(+0,t) - \lambda_b^L n_b^L$$

$$\lambda_b^L n_b^L = c_b^L \rho_b(+0,t) - j_b(+0,t)$$

where $j_a(x,t)$, $n_a^L(t)$ and $\rho_a(x,t)$, $j_b(x,t)$, $n_b^L(t)$ and $\rho_b(x,t)$ are respectively fluxes, surface and volume concentrations of complex $(KB)^+$ (particle a) and the neutral complex-formator B (particle b); α_{Kb}^L and μ_a^L are the rates of the complex-formation process and the complex decomposition.

Boundary conditions at $x = d$ can be written similarily and then assuming thermodynamic equilibrium the following equality should be satisfied

$$\frac{\alpha_{KB}^{L(R)} c_b^{L(R)}}{\lambda_b^{L(R)}} a_K^{L(R)} \rho_b[0(d)] = \frac{\mu_a^L c_a^L}{\lambda_a^L} \rho_a[0(d)] \tag{5.8}$$

Using the found boundary conditions (their generalization for cases of other chemical transformations on the surface are evident), one can consider processes of ion and molecule transfer through membranes. Paper [236] discusses processes of a conjugated counter (antiport) and a unidirected (symport) ion transfer through membranes using boundary conditions of this type. The paper gives expressions for the rate of ion transfer, including the case of active transport (ion movement along the concentration gradient) and the corresponding values of membrane potentials if nonequilibrium effects in boundary conditions are taken into account.

Also note that the use of boundary conditions (5.1–5.8) makes unnecessary the unfounded introduction of such membrane characteristics as anti-ion and co-ion electric current independent transfer numbers in the membrane. It also allows considering problems of transmembrane ion transfer with an arbitrary relation between the rates of surface chemical processes and the rate of diffusion particle motion in the membrane (unlike the limit situations of relatively high and low rates of surface processes considered in [232] using expansions in the Dömkoler model parameter).

5.1.2. Ion Transfer Specificity in the Interphase Boundary Region

As has been mentioned above, the kinetics of ion transfer through the interphase boundary (and consequently, parameters λ, μ, c appearing in the corresponding boundary conditions) can to a great extent be determined by the specificity of the interphase boundary state—the inevitable presence in the phase conjugation region of rather strong electric fields and considerable structural nonuniformities in the boundary region. Let us analyze both these factors in more detail.

5.1.2.1. Electric Submicroheterogeneity of Boundaries

We assumed above that the interphase boundary electrolyte-membrane is characterized by a certain (equal for all points of the interface) value of the electric potential jump and that the parameters of the boundary conditions depend on this value. However, the ions passing through the interphase region can be influenced by strong inhomogeneous (random) electric fields produced by various dipole fragments localized in the region of phase conjugation. Such fields may appreciably effect the flux of ions passing through the interphase boundary even at small degrees θ of surface occupation, $\theta \lesssim 0.1$.

The corresponding estimates are taken into consideration in paper [233] which analyzes the effect of electrostatic fields from adsorbed molecules of surfactants upon the flux of charged particles through the interphase boundary. If dipole molecules are adsorbed with surface concentration N_S on the interphase boundary so that one end of the dipole with charge qe is in a solution with dielectric permittivity ε at distance l_0 from the interphase boundary localizing the opposite charge— $-qe$, then the electric fields created (at $q > 0$) suppress the K cation contribution to the flux through the membrane into the solution and the insertion of A anions from the solution into the membrane. The relative decrease of the corresponding fluxes at random distribution of surfactant molecules over the surface at low occupation degrees ($\theta \lesssim 0.1$) is equal to [233]:

$$\frac{\bar{j}_i}{j_{i,0}} = \exp\left\{-\pi\gamma^{2/3} N_S \left[\Gamma\left(\frac{1}{3}\right) - \frac{a_i^2}{\gamma^{2/3}} + \frac{2}{3}\,\Gamma\left(-\frac{2}{3}, \frac{\gamma}{a_i^3}\right)\right]\right\},$$

$$\gamma = \frac{2l_0^2\alpha}{\varepsilon KT}\,|qq_0|, \qquad a_a = 2l_0, \qquad a_K = l_0 \tag{5.9}$$

where α is the phenomenological parameter of the potential barrier (the characteristic width that should be overcome by a particle with charge

q_0e passing over the interphase boundary); $\Gamma(z)$ and $\Gamma(x, z)$ are respectively the gamma function and the incomplete gamma function.

It can be easily proved that at small values of l_0 when $l_0^2/\gamma^{2/3} \ll 1$ Equation (5.9) becomes expression

$$\bar{j}_i / j_{i,0} = \exp\left[-\pi\Gamma\left(\frac{1}{3}\right)\gamma^{2/3}N_s\right] \tag{5.10}$$

while at $l_0^2/\gamma^{2/3} \gg 1$ Equation (5.9) gives the result that is obtained from the model of delocalized charges usually employed (the expression for field intensity of the plane condenser):

$$\bar{j}_i / j_{i,0} = \exp[-(2\pi\gamma/a_i)N_s] \tag{5.11}$$

Note the nontriviality of result (5.10) which predicts the non-Arrhenius character of flux $\bar{j}_K$ suppression upon dipole molecule absorption with $q > 0$. A comparison of (5.10) and (5.11) shows that taking into account the discreteness of charges on the membrane surface is absolutely necessary for calculating field effects (if we choose $\alpha = 0.1$ nm, $|q| \approx |q_0| \approx 1$, $T = 300$ K, inequality $l_0 < \gamma^{1/3}$ is violated only at $l_0 >$ $> (0.4 - 1.6)$ nm when $\varepsilon = 80 - 5$ accordingly. It is the field dependencies (5.10) which are characteristic for the case when charges qe cannot be considered delocalized in the corresponding planes that may lead to a considerable (by several orders) $\bar{j}_i$ current suppression even at a low degree of surface occupation, $\theta \lesssim 0.1$.

Apparently taking into account surface electric inhomogeneity effects may appear essential when investigating the influence of surface chemical treatment of ion-exchange membranes (e.g., at absorption of surfactants [237–239]) upon membrane selectivity. Electric fields in the region of interphase boundaries may appreciably change rate values μ, c and λ of elementary process in the case of mass transfer through the charged particle interface. This opens up possibilities for regulating the surface states (e.g., by adsorption) to control mass transfer through the interphase boundaries. Recently, [239] such a possibility has been analyzed for ion transfer in biological membranes (see [240]) and field dependencies similar to (5.9–5.10) have be obtained.

In conclusion, it should be said that traditionally the kinetics of ion transfer through the interphase boundary ion exchange membrane—electrolyte solution is less studied than ion transport in the membrane volume. Nevertheless, we think it possible to obtain the desired changes in the physical-chemical parameters of membranes, and not only ion-exchange ones. This can be done by standardizing the surface state of ion-exchange membranes in contact with electrolytes of different concentration (at high concentrations the structure of the solution may considerably influence interphase boundary parameters) and a controlled modification of subthin surface layers (see Section 5.3).

An investigation of the surface state of ion-exchange membranes also provides a basis for a systematic search for the most effective nonstationary membrane processes. In particular, alternating currents can be used in industrial processes with ion-exchange membranes (electrolysis, electrodialysis, extraction).

5.1.2.2. Ion Generation

Strong electric fields in the interface region cannot only change considerably ion flux but also stipulate the limit nonequilibrium case—ion generation in these regions in the process of heterolytic molecule dissociation. Let us, as an example, consider the process of heterolytic dissociation of water molecules proceeding in bipolar ion-exchange membranes in the spatial charge region on the boundary of a cation and anion membrane [240–242]. The dissociation process of water molecules as has been shown in [243] depends to a large extent on the type of ionogenic groups in the bipolar membrane (it appeared that the most effective are the weakly-acid group, the least effective the strongly-acid ones). This fact is determined by the specificity of water molecule coordination near the ionogenic group and by the polarizing influence of the Coulomb field of the group upon water molecules [244]. The potential energy term of the system, water molecule—ionogenic group may considerably deform the same as its vibration characteristics along the coordinate corresponding to the dissociated decay of water molecules. Therefore the initial rate constant k_0 of water molecule dissociation in this case may noticeably exceed the corresponding rate constant for free water molecules.

According to [242] the field dependence of the rate constant k_F of water molecule electrolytic dissociation in an electric field of strength F has the form

$$k_F = k_0 \exp(\alpha F) \tag{5.12}$$

where α is a factor which weakly changes with temperature (entropy) the value of which depends on the type of functional group which may vary in the interval: $\alpha = (3.0 - 4.5)\ 10^{-9}$ m/V.

Ions H^+ and OH^- resulting from the elementary act of dissociation are carried out by a strong electric field outside the boundaries of the spatial charge region (in this case current density is determined entirely by the rate of H^+ and OH^- ion generation in elementary processes).

An analysis of membrane MB-1, MB-2 volt-ampere characteristics in the large current region in the temperature interval from 25 up to 55°C upon changing the concentration of the outer electrolyte (solution NaCl, as well as solutions NaOH from the side of the anionite membrane and HCl from the side of the cationite membrane) from 0.01 to 0.5 M, showed [230] the adequacy of expression (5.12).

The electrolyte dissociation of water molecules yielding H^+ and OH^- ions is one of the stages of many electrochemical processes: salt solution electrolysis, water decomposition producing hydrogen and oxygen, corrosion and oxidation processes, biological processes at the subcellular level. The established dependence (5.12) may be applied for analyzing these processes.

5.1.3. Nonequilibrium Effects on the Ion-Exchange Membrane—Electrolyte Solution Boundaries

Let us consider the simplest examples illustrating the use of the boundary conditions introduced above for describing ion transport and the kinetics of electrodiffusion processes in systems of ion-exchange membranes.

5.1.3.1. *Conjugated Ion Transfer*

The ion flux through the ion-exchange membrane may be appreciably increased (up to active transport—the movement of ions on the concentration gradient) [236] if we create a conjugated—counter (antiport) or a unidirected (symport) ion flux of the same or opposite sign respectively through the membrane. Our interest lies with finding the dependencies of the flux density of transferred ions on ion concentrations (activities) in the electrolyte on both sides of the membrane; dependencies on the process parameters of ion insertion into the membrane on the electrolyte-membrane boundaries; and on density J of the summary electric current in the system.

We shall confine ourselves to considering the simplest stationary case of the kinetics of single-charged ion transfer (e.g., Ag^+) from the extraction chamber (L- chamber) through an ideal cation-exchange membrane (assuming that co-ions, e.g., NO_2 are not included in the membrane) with concentration ρ_0 of the ionogenic groups in the reextraction chamber (R- chamber) at a counter transfer of more mobile single charged ions (H^+) under the condition of external polarization.

We shall use the most general type boundary conditions depending on fluxes J_1 and J_2 of transferred ions and protons, respectively, and electric potential jumps $\psi_D^{L(R)} = e/kT[\varphi_{L(R)} - \bar{\varphi}_{L(R)}]$ in the L and R region of interphase boundaries membrane—solution ($\varphi_{L(R)}$ and $\bar{\varphi}_{L(R)}$ are the $L(R)$ boundary potentials from the side of the electrolyte and the membrane; we ignore the influence of $\psi'_{L(R)}$—potentials on the kinetics of ion-transfer into the membrane from the side of the electrolyte assuming that electrolyte solutions in L- and R- chambers are sufficiently concentrated). Considering the system of boundary conditions jointly with the electroneutrality condition of ion-exchange membrane volume, and with the expressions for ion fluxes at a given total current density $J = J_1 + J_2$, enables us to derive a system of algebraic equations for determining the sought after dependencies $J_1(J)$ and $\psi_M(J) = e/kT(\varphi_L - \varphi_R)$ of the extracted ion flux and reduced membrane potential, respectively:

$$\gamma_0^{(1)}\left(qf - \frac{D_1}{D_2 - D_1}J\right) + x_{00}^{(1)}L_0(B + C) = \alpha_L^{(1)}a_L^{(1)}$$

$$\gamma_d^{(1)}\left(-qf + \frac{D_1}{D_2 - D_1}J\right) + x_{d0}^{(1)}L_d[B + C\exp(fd)] = \alpha_R^{(1)}a_R^{(1)}$$

$$\gamma_0^{(2)}\left(-qf + \frac{D_2}{D_2 - D_1}J\right) + x_{00}^{(2)}L_0[\rho_0 - B - C] = \alpha_L^{(2)}a_L^{(2)} \tag{5.13}$$

$$\gamma_d^{(2)}\left(qf - \frac{D_2}{D_2 - D_1}J\right) + x_{d0}^{(2)}L_d[\rho_0 - B - C\exp(fd)] = \alpha_R^{(2)}a_R^{(2)}$$

Here

$$f = -\frac{e}{kT}\frac{d\bar{\varphi}(x)}{dx}, \qquad L_{0(d)} = \exp[-\psi_D^{L(R)}] \tag{5.14}$$

$$q = \frac{e\rho_0 D_1 D_2}{D_2 - D_1}, \qquad B = \frac{D_2}{D_2 - D_1}\rho_0 - \frac{J}{ef(D_2 - D_1)} \tag{5.15}$$

where $\bar{\varphi}(x)$ is the local value of the potential in the membrane (x is the coordinate, normal to the membrane surface, $0 \leqslant x \leqslant d$, d is the membrane width); D_1 and D_2 are, respectively, the ion and proton diffusion coefficients ($D_2 > D_1$).

Solving the system (5.13) with respect to f, L_0, L_d and C one can find distributions $\rho_1(x)$ and $\rho_2(x)$ in the width of ion and proton concentrations respectively in the membrane, membrane potential ψ_M and its components, fluxes J_1 and J_2:

$$\rho_1(x) = B + C\exp(fx), \qquad \rho_2 = \rho_0 - \rho_1 \tag{5.16}$$

$$\psi_M = \ln\frac{L_d}{L_0} + fd \tag{5.17}$$

$$J_1 = J_0 - \frac{D_1}{D_2 - D_1}J, \qquad J_2 = -J_0 + \frac{D_2}{D_2 - D_1} \tag{5.18}$$

where $J_0 = qf$ is the exchange current.

In the particular case $\gamma_{0(d)}^{(1,2)} = \gamma_{0(d)}$, $x_{0(d)0}^{(1,2)} = x_{0(d)}$, $\xi = 0$ one has

$$L_0 = \frac{1}{x_0\rho_0}(\alpha_L^{(1)}a_L^{(1)} + \alpha_L^{(2)}a_L^{(2)} + \gamma_0 J) \tag{5.19}$$

$$L_d = \frac{1}{x_d\rho_0}(\alpha_R^{(1)}a_R^{(1)} + \alpha_R^{(2)}a_R^{(2)} + \gamma_d J) \tag{5.20}$$

Value $f[a_{L(R)}^{(1,2)};\ \gamma_{0(d)};\ x_{0(d)};\ D_1, D_2]$ can be found by solving the transcendental equation:

$$\frac{1}{x_0L_0}(\alpha_L^{(1)}a_L^{(1)} - \gamma_0J_1) + \frac{1}{x_dL_d}(\alpha_R^{(2)}a_R^{(2)} + \gamma_dJ_2)\exp(-fd) =$$

$$= B + (\rho_0 - B)\exp(-fd) \tag{5.21}$$

The system of Equations (5.13–5.15) and dependencies (5.16–5.20) can be used in the experimental investigations of the kinetics of ion extraction through the polymer ion-exchange membrane. An analysis of dependencies $J_1(J)$ and $\psi_M(J)$ and different membrane widths and values of $a_{L(R)}^{(1,2)}$ permits determining the kinetic parameters of the investigated system (diffusion coefficients, boundary condition parameters). Such an analysis provides for controling the state of the interphase boundary membrane-solution and, in particular, those changes in the surface state which result from various thermal treatments of the membranes (surfactant adsorption, chemical and plasmochemical surface modifications).

Similarily, one can consider more complicated models of ion transfer including transfer with the participation of complex-formators. Difficulties arise when describing the transfer kinetics in liquid extraction membranes, where conditions for the appearance of free convections may arise.

5.1.3.2. *Kinetics of Ion Transfer in an Electromembrane System*

Let us consider [245] the stationary process of electrodiffusion in a system comprising one horizontally arranged cationic membrane separating the stirred solutions of 1,1-valent electrolyte. The anti-ion current is directed from bottom to top which ensures the absence of natural convection in the immobile layer of δ width adjoining the membrane. Concentrations in the stirred regions of the solution remain constant in the electrodiffusion process which can be obtained by an intense stirring in sufficiently large solution volume. It is assumed that all the conditions ensuring the one-dimensionality of electromass-transfer in the direction perpendicular to the membrane surface of d width are satisfied. The density of J current directed perpendicularly to the membrane surface is constant throughout the process. Coordinate x along transport direction of cations 1 and anions 2 originates in the middle of the membrane, its positive direction coinciding with the direction of the current of the cation flux. The chamber concentration is denoted by R and that of demineralization—by L.

The conditions on the boundary membrane-solution are balance equations for ion fluxes between surface state $I_{L(R)}$ and boundary states in the solution $II_{L(R)}^{(e)}$ and in the membrane II_{LR}. In the case of a stationary ion transfer on the boundary I_L, one obtaines the following condition:

$$\alpha_i^L \rho_i \left(-\frac{d}{2}\right) - \left(\lambda_i^L + \mu_i^L\right) n_i^L + c_i^{L-} \bar{\rho}_i \left(-\frac{d}{2}\right) = 0, \quad i = 1,2 \tag{5.22}$$

where $\rho_i(-d/2)$ is the concentration of i-type ions in a solution at the membrane surface, i.e., at $x = -d/2$, $\bar{\rho}_i(-d/2)$ is the boundary concentration of the same ions in the membrane.

Balance equations on the boundary for the case of single-charged ions of two types have the form

$$c_1^{L-} \bar{\rho}_1 \left(-\frac{d}{2}\right) + j_1 = \lambda_1^L n_1^L \qquad c_2^{L-} \bar{\rho}_2 \left(-\frac{d}{2}\right) - j_2 = \lambda_2^L n_2^L \tag{5.23}$$

where j_1 and j_2 are the densities of cation and anion fluxes.

From (5.22–5.23) one can obtain the conditions on the membrane boundary with L-region of the electrolyte.

$$\rho_1\left(-\frac{d}{2}\right)\alpha_{1,0}^{\ L} = x_{1,0}^{\ L} \exp(-\psi_L)\bar{\rho}_1\left(-\frac{d}{2}\right) + (1+\gamma_{1,0}^{\ L})j_1 \qquad (5.24)$$

$$\rho_2\left(-\frac{d}{2}\right)\alpha_{2,0}^{\ L} = x_{2,0}^{\ L} \exp(\psi_L)\bar{\rho}_2\left(-\frac{d}{2}\right) - (1+\gamma_{2,0}^{\ L})j_2$$

where $x_{i,0}^{\ L} = \mu_{i,0}^{\ L} c_{i,0}^{\ L}/\lambda_{i,0}^{\ L}$; $\gamma_{i,0}^{\ L} = \mu_{i,0}^{\ L}/\lambda_{i,0}^{\ L}$; ψ_L is the dimensionless jump of the electric potential in the region of the spatial membrane charge on the condition that rates μ_i and α_i are independent of potential jumps in the electrolyte subsurface region and $\xi_i^L = 0$.

Similarily one obtains for the membrane boundary with R-region

$$\rho_1\left(\frac{d}{2}\right)\alpha_{1,0}^{\ R} = x_{1,0}^{\ R} \exp(-\psi_R)\bar{\rho}_1\left(\frac{d}{2}\right) - (1+\gamma_{1,0}^{\ R})j_1 \qquad (5.25)$$

$$\rho_2\left(\frac{d}{2}\right)\alpha_{2,0}^{\ R} = x_{2,0}^{\ R} \exp(\psi_R)\bar{\rho}_2\left(\frac{d}{2}\right) + (1+\gamma_{2,0}^{\ R})j_2$$

Below, the initial rates of elementary stages for both boundaries of the membrane and solution will be considered to be equal (in neglecting potential jumps).

Assuming the electroneutrality of the solution in δ-layers and the membrane, when considering stationary electromass-transfer in the direction of coordinate x for $-(d/2+\delta_L) \leqslant x \leqslant d/2$, we have

$$j_1 = -D_1\frac{d\rho_1}{dx} + D_1 a_L \rho_1, \qquad j_2 = D_2\frac{d\rho_2}{dx} + D_2 a_L \rho_2 \qquad (5.26)$$

$$j = j_1 + j_2 = \frac{J}{e}$$

$$\rho_1 = \rho_2 \qquad \rho_{1(2)}\left(-\frac{d}{2} - \delta_L\right) = \rho_L^{\ 0}$$

where $a_L = -d\psi/dx$ is the electric field strength; D_1, D_2 are the diffusion coefficients of cations and anions in the solution.

Writing equations analogous to (5.26) for the δ_R layer and the membrane and solving them by applying boundary conditions (5.24–5.25), one can determine concentration distribution and the electric potential in δ-layers and the membrane as well as ion fluxes. The simplest form of the problem's solution has at low intensities of the electric field in the solution and membrane: $a_{L(R)}\delta_{L(R)} < 1$, $\bar{a}d < 1$ at equal initial concentrations and width of diffusion layers on both sides of the membrane $\rho_L^{\ 0} = \rho_R^{\ 0} = \rho^0$, $\delta_L = \delta_R = \delta$, equal cation and anion diffusion coefficients $D_1 = D_2 = D$, $\bar{D}_1 = \bar{D}_2 = \bar{D}$ and a sufficiently high concentration of fixed groups in the membrane, $\bar{\rho^0} > \rho^0$:

$$j_{1(2)} = J/2e, \qquad a_L = -a_R = a = j/\rho^0 D, \qquad \bar{a} = j_1/\overline{\rho^0}\,\bar{D}$$

$$\psi_L = \ln \frac{\overline{\rho^0} x_{1,0}}{\rho^0(1 + a\delta)\alpha_{1,0} - \left[\frac{\delta}{D}\,\alpha_{1,0} + (1 + \gamma_{1,0})\right] j_1} \tag{5.27}$$

$$\psi_R = \ln \frac{\overline{\rho^0} x_{1,0}}{\rho^0(1 + a\delta)\alpha_{1,0} + \left[\frac{\delta}{D}\,\alpha_{1,0} + (1 + \gamma_{1,0})\right] j_1}$$

$$\rho_1 = \rho^0 + (\rho^0 a) - j_1/D)[x \pm (d/2 + \delta)]$$

In the last equation the "+" sign corresponds to the demineralization chamber, sign "−" to the concentration chamber. The dimensionless membrane potential has the form

$$\Delta\psi_M = -\bar{a}d + \psi_R - \psi_L \tag{5.28}$$

Figure 5.1 presents the dependencies of the dimensionless membrane potential and its components on varying quantities $\alpha_{1,0}$, $x_{1,0}$ and $\gamma_{1,0}$ at

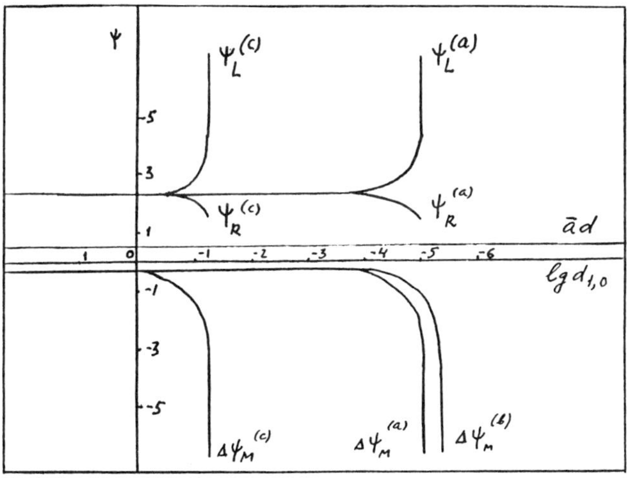

Fig. 5.1. Dependence of the dimensionless membrane potential and its components on boundary condition parameters. The curves correspond to $J = 10^{-1}$ mA/cm^2, $\delta = 10^{-2}$ cm, $d = 5 \cdot 10^{-2}$ cm, $D = 10^{-5}$ cm^2/s, $\bar{D} = 10^{-6}$ cm^2/s, $\rho^0/N_A = 10^{-1}$ M, $\bar{\rho}^0/N_A = 1$ M (N_A is the Avagadro number)

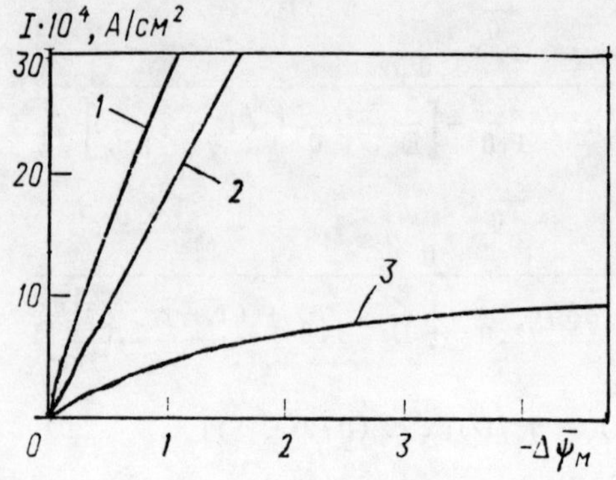

Fig. 5.2. Dependence of the dimensionless membrane potential on current density: 1 – $d = (0.1 - 10^3)$ cm/s; 2 – $d = 10^{-3}$ cm/s; 3 – $d = 10^{-4}$ cm/s.

definite values of parameters characterizing the properties of the system and the regime of the process. Case '*a*' corresponds to equal rates of anti-ion transition from the adsorbed state into bulk states both for the membrane and the solution ($\alpha_{1,0} = c_{1,0}$) $\mu_{1,0} = \lambda_{1,0}$); case '*b*'—to the anti-ion transfer dominant from the solution into the membrane ($\alpha_{1,0} > > \mu_{1,0} < \lambda_{1,0} > c_{1,0}$); case '*c*'—vice versa ($\alpha_{1,0} < \mu_{1,0} > \lambda_{1,0} < c_{1,0}$).

Figure 5.2 illustrates the dependencies (calculated at $\alpha_{1,0} = c_{1,0}$; $\mu_{1,0} = \lambda_{1,0}$) of the dimensionless membrane potential on current density.

According to data in Figure 5.1 and Figure 5.2 the membrane potential is independent of the values of kinetic parameters at sufficiently high rates of the elementary stages of ion transport. Volt-ampere characteristics of the electromembrane system (neglecting the drop in potential in δ layers) are linear and practically coincide which agrees with the notions of local thermodynamic equilibrium on the boundary membrane—solution [246]. A rate decrease leads to a dramatic increase of the membrane potential, and consequently to "locking" the membrane at sufficiently low densities of the electric current passing through the system, i.e., to revealing the nonequilibrium character of ion transfer through the boundary membrane-solution.

If this approach is extended to consider ion electrodiffusion in membrane systems, it becomes possible to adequately describe various nonstationary processes and to trace the influence of the state of the boundary membrane-solution on ion transport efficiency.

5.1.4. Influence of the Structural Specificity of Polymer Ion-Exchange Membranes on Ion Transfer

The kinetics of ion transfer in the ion-exchange membrane volume is, to a great extent, determined by the structural specificity of the polymer matrix. The most characteristic property of the structure of such membranes is nonuniformity at the submicroscopic level of ion group distribution. Let us briefly discuss ion transfer processes through new ion-exchange materials—perfluorinated cationic membranes (sulphacationic membranes "Nafion" and MF 4CK, membranes with carboxyl ionogenic groups) which due to their chemical stability are applied in various fields of technology and industry, in particular, for the electrolytic production of high-quality chlorine and caustic soda.

Physical chemical and structural investigations of perfluorinated membranes performed in various laboratories (the methods used include IR spectroscopy [247,248], X-ray scattering using small and large angles [249,250], Mössbauer spectroscopy [251], ESR and NMR relaxation [252, 253], high resolution NMR, double electron-nuclear resonance [254], standard poremetry [255], and kinetic methods for studying the transfer of various ions and water molecules [249,256,257]) have established the fact of ionogenic group aggregation in membranes with the formation of clusters. Such aggregation considerably influences both the course of ion and water molecule transfer through the membrane and the mechanical properties of the membranes.

Let us present the submicrostructure of a perfluorinated ion-exchange membrane following paper [250], where a conclusion was drawn on the crest-like character of the arrangement of the co-polymer side branches. This conclusion was made on the basis of investigating the small-angle diffractograms of the unhydrolyzed copolymer and the membrane in the isotropic and uniaxially-orientated states. Figure 5.3 presents a fragment of the amorphous part of the membrane, schematically showing the hydrated ion-exchange groups between which ions jump in the course of their transport through the membrane.

According to reference data (see [257]), the basic laws of ion transfer are, to a great extent, determined by specific membrane moisture capacity 'n', the number of water molecules per ionogenic group, which noticeably decreases with a concentration increase of the external electrolyte. So, value 'n' dependencies of Na^+ ion self-diffusion coefficients $D_{Na}(n)$ of OH^-, Cl^- co-ion diffusion coefficients $D_{OH}(n)$ and $D_{Cl}(n)$ in sulphacationite membranes MF—4 CK have a threshold character and at relatively small 'n' values ($n \lesssim 10$) may be approximated in the following way [257]:

$$D_K = A_K \exp\left[-p_K(a_K - n^{1/3})^{\mu_K}\right] - \left[-\frac{E_K(n)}{kT}\right] \tag{5.29}$$

The values of the corresponding parameters are given in Table 5.1.

Paper [257] suggested a microscopic model based on the notions of percolation theory (of liquid in a random labyrinth) for describing the kinetic laws of ion transfer in a perfluorinated membrane matrix and for

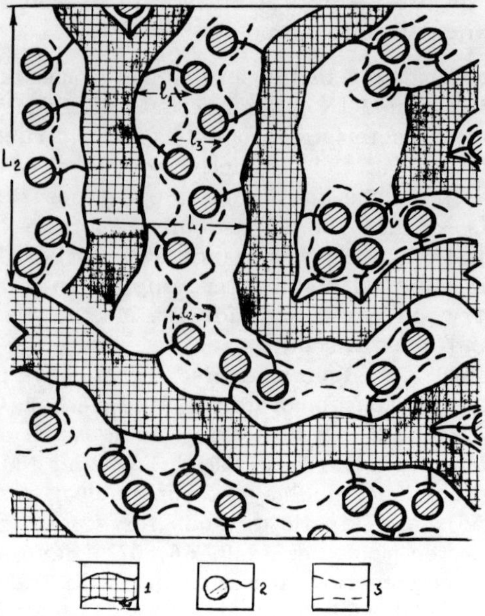

Fig. 5.3. Schematic fragment of the amorphous part of the perfluorinated ion-exchange membrane. The main fluorine-carbon polymer chain is shaded.

calculating transfer coefficients (of diffusion). The ion trajectory in this case is random chains (infinite clusters) passing through the whole membrane comprised of connected sites where the ions may localize. The probabilities of ion jumps (the coupling between the adjacent sites of a random net of ion trajectories) depend not only on mean distance $\overline{R}$ between associates ($\overline{R} \approx 1$ nm) but on the degree of their hydration. An increase in the hydration degree increases the probability for the formation

Table 5.1. Values of parameters A_K, p_K, a_K, μ_K (index K denotes the type of ion: Na^+, OH^- or Cl^-) for two membrane series MF-4CK with different ion exchange capacities.

K	Q = 1.12 mole eq/l				Q = 1.23 mole eq/l			
	A_K, m^2/s	p_K	q_K	μ_K	A_K, m^2/s	p_K	q_K	μ_K
Na^+	$1.46 \cdot 10^3$	23	3.15	0.42	3.9	9.3	3.32	1.88
OH^-	1.66	17.3	3.17	0.28	$5.84 \cdot 10^2$	22.7	3.32	0.23
Cl^-	$5.6 \cdot 10^2$	22.2	3.17	0.43	1.0	15.0	3.32	0.51

of bridge bonds (over water molecules) between associates and consequently the probability of ion jumps between them.

Let us represent probability $w_{ij}^{(K)}$ of a K ion jump between hydration associates (sites) i and j separated by distance R_{ij} in the following form:

$$w_{ij}^{(K)} = w_0^{(K)} \exp\left(-\xi_{ij}^{(K)}\right) \tag{5.30}$$

$$\xi_{ij}^{(K)} = \frac{\varepsilon_{ij}}{kT} + \{\alpha_K [R_{ij} - r_i(n_i) - r_j(n_j)]\}^{\mu_K} \tag{5.31}$$

where ε_{ij} is the activation energy for a jump of ion K between sites i and j; $r_i(n_i)$ is the radius of the i-th associate, containing n_i water molecules; the second term in exponent (5.31) reflects the coupling of adjacent sites for K ions; α_K and μ_K are the phenomenological parameters of this bond.

We assume two sites to be coupled (an ion transition is possible between them) if the following relation is valid

$$\xi_{ij} \leqslant \xi \tag{5.32}$$

The percolation theory problem being sought consists of finding threshold value $\xi = \xi_c$, when the coupled sites form infinite paths (infinite clusters) [258].

If the sites of the trajectory net are randomly (uncorrelatedly) arranged over the volume (N is their average concentration) probability d^3f for finding the nearest one (from some given site in elementary volume $dRd^2\Omega$) in direction Ω at distance R is equal to $d^3f = R^2N \exp(-1/3 \times \times NR^3)dRd^2\Omega$. In this case, the solution of the percolation theory problem on the condition of equal hydration of all centers $[r_i(n_i) = r(n)]$ gives for the transfer coefficient dependence (5.29) at the following parameter expressions:

$$P_K = \left[2\alpha_K \left(\frac{3}{4\pi\rho_0}\right)^{1/3}\right]^{\mu_K},$$

$$a_K^3 \equiv \bar{n} = \frac{\pi\rho_0}{2N} \ln^3\left(1 - \frac{\nu_c}{4r}\right)^{-1} \tag{5.33}$$

where ρ_0 is the effective density of water molecules in hydrated associates; $\nu_c = (3.0 \pm 0.1)$ is the percolation theory parameter (the average number of bonds of a separate center required for the appearance of infinite clusters in the system), the numerical value of ν_c is taken from paper [259].

A comparative analysis of the data from Table 5.1 and relations (5.33) shows that the model expression (5.30) considered for the ion jump probability as well as the assumption about the random distribution of hydrated associates over the matrix, qualitatively describe the observed regularities.

In particular, this is proved by the weak dependence of parameter a_K on the type of ions for all the investigated membranes (note, that in the case of a membrane with exchange capacity $Q = 0.78$ value a_K is $a_K = 3.18$). The decrease of a_K when passing to membranes in Cs^+ form points to the less effective density ρ_0 of water molecules upon ionogenic group hydration in this case. However, the difference in P_K values of different membranes testifies to the simplified character of the considered model.

It follows from relation (5.29) and Table 5.1 that diffusion coefficients (D_{Na}, D_{OH}, D_{Cl}) are of the same order, the coupling in co-ions between associates being realized even at a lesser hydration than in the case of Na^+ anti-ion transfer for all the investigated membranes (at small n values Na^+ ions are not transferred through the membrane because of strong coupling with ionogenic groups). However the penetrability of sulphacationic perfluorinated membranes over co-ions OH^-, Cl^- can be considerably lowered by chemical modification (in particular, amination [260]) of perfluorinated membranes.

Such modification permits sharply decreasing the distribution coefficient γ_K equal to the ratio of the electrolyte solution concentrations in the membrane and outside it (diffusion penetrability of the membrane with respect to co-ions is determined by the product $\gamma_K D_K$).

Indeed, in nonmodified membranes MF-4CK, distribution coefficients of co-ions OH^- in alkali solutions (6 – 16 M) and co-ions Cl^- in the NaCl solution (5 M) are equal: $\gamma_{OH} = 0.3 - 0.4$, $\gamma_{Cl} = 0.17$ for membranes with $Q = 0.78$ mole eq/l; $\gamma_{OH} = 0.6 - 1.0$, $\gamma_{Cl} = 0.15 - 0.17$ for membranes with $Q = 1.12$ and 1.23 mole eq/l. In ethylenediamine membranes aminated throughout the entire volume the value of co-ion distribution coefficients are much smaller (especially in a highly concentrated alkali (Figure 5.4))

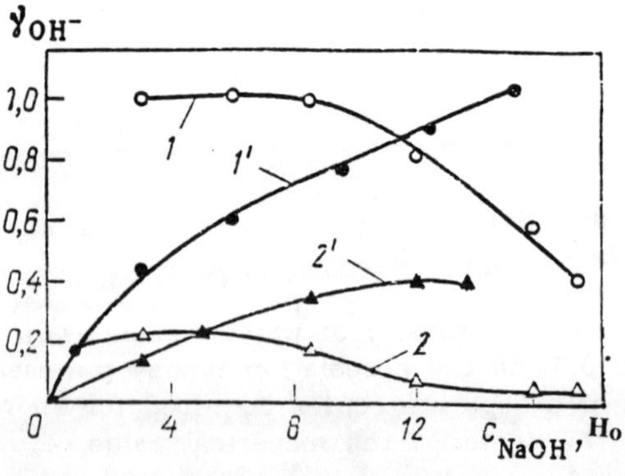

Fig. 5.4. Dependence of distribution coefficients on the concentration of external solutions for membranes with exchange capacities $Q = 0.78$ (1,1') and 1.23 Mole Eq/l (2,2'): 1, 2 (and 1',2') corresponding to fully-aminated (unaminated) membranes.

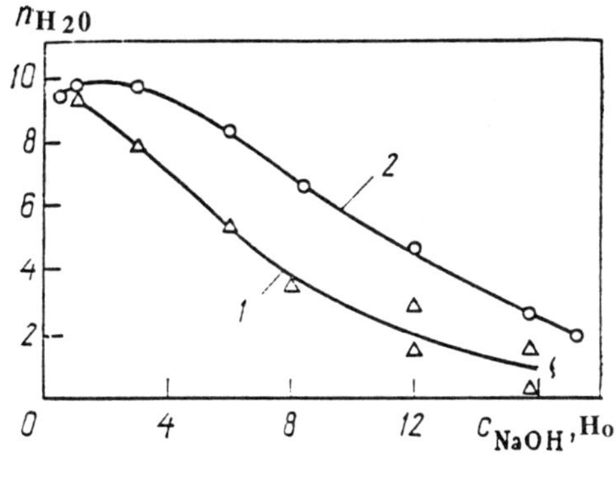

Fig. 5.5. Dependence of membrane moisture capacity n per one ionogenic group on concentration of outer NaOH solutions for aminated membranes with Q = 0.78 (1) and 1.23 Mole Eq/l (2).

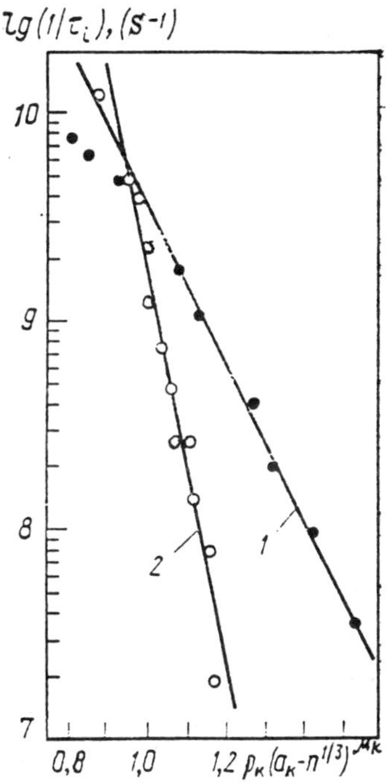

Fig. 5.6. Experimental and calculated dependencies of water molecule rotary mobility (1) and proton exchange frequency (2) on moisture content for MF-4CK membranes the in H^+—form. The calculation is according to (5.29) at a = 3.17, p = 9.2, μ = 0.5 (1) and a = 3.17, p =22, μ = 0.2 (2).

due to a noticeable decrease in membrane swelling and moisture content (Figure 5.5). Progress in investigating transfer processes in ion-exchange membranes is connected with the wide application of new methods for studying the membrane submicrostructure and the dynamics of molecular motion in them. Important information on the parameters and the character of water molecule and proton motion in the liquid phase of the membrane can be provided by radiospectroscopy methods: the application of the NMR relaxation method [253] established the dependence of proton exchange frequency τ_1^{-1} and the water molecule rotary motion τ_2^{-1} on value n. It appears that these dependencies are well approximated by the same expressions of type (5.29) as dependence $D_K(n)$ (Figure 5.6). This fact, as well as the coincidence observed of values D_{H_2O} and $D_{H_5O_2^+}$ obtained by the NMR method and macrotransfer data, prove that at $n < 10$ hydrate sulpha-group shells are not overlapped inside the clusters. The hydrate shells of ionogenic groups include, besides the functional groups and water molecules, molecules of the inexchangeably absorbed electrolyte and anti-ions. In all probability, the molecules in such hydrate shells are structured the same as in the case of subthin layers of liquids ($\lesssim$ 1 nm near hydrophilic surfaces [261]. Hence the particle transfer from one ionogenic group to another is connected not only with overcoming the energy barriers but also with the structure reconstruction of hydrated ion complexes resulting from packing fluctuations. This fact is reflected in the presence of an entropy multiplier in (5.29) which characterizes the moisture content dependence of molecular and ion mobility in the liquid membrane phase.

The concepts developed on ion transfer processes in perfluorinated sulphacationic membranes lead to important conclusions on the possibility of controlling ion transfer under the condition of sodium salt electrolysis. To increase the selectivity of anti-ion transfer without a noticeable increase in membrane electric resistance, it may prove sufficient to create in presurface membrane regions subthin layers with less moisture content which prohibit co-ion transport. In perfluorinated membranes the macroscopic surface layers (tens of micrometers) are usually modified. This is achieved not only by chemical treatment but also by lowering ionogenic group concentration in presurface layers upon thermal treatment, by making sandwich membranes out of polymers with different equivalent mass, etc. However, the analysis performed shows that the modified surface layer may be considerably (by two orders) thinner, i.e., not exceed L = 100 nm if the layer possesses sufficiently homogeneous properties. Such film thicknesses L are already "infinite" ($L \gg R$) according to the notions of percolation theory. Therefore the conditions for the appearance (or nonappearance if co-ion transfer is meant) of connected ion trajectories (infinite clusters in terms of the percolation theory), passing through a layer of the given thickness, are the same as for macroscopic layers with $L \approx 10^{-2}$ cm. These conditions are defined only by values Q and n and are independent of layer thickness. Therefore an increase in the thickness of the selective layer over $L \approx 100$ nm leads only to an undesired increase in membrane resistance, and possibly to worsening the adhesion of the modified layer with the unmodified polymer.

5.2. WATER AND ION TRANSFER IN REVERSE OSMOTIC MEMBRANES

The phenomenon of reverse osmotic (RO) water demineralization using semi-penetrable membranes (penetrable for water molecules—the solvent and practically nonpenetrable for hydrated salt ions) upon applying pressures higher than the osmotic one is widely used in water preparation processes. We shall not touch upon the question connected with the organization of such a process (for information about the types of membrane elements and apparatus schemes of RO-devices see [214–216; 220–223]). Our principal aim here is to discuss important and yet unclear questions on the physics and chemistry of reverse osmotic membrane separation of liquid mixtures, i.e., the RO-separation mechanism, and the role of interphase boundaries in RO processes.

RO-membranes are usually used in the form of asymmetric cellulose acetate films, in which only the surface layer ($\lesssim$ 1 mcm), the so-called selective or active layer containing submicropores with characteristic size $R \lesssim 1$ nm, [262] possesses separation capacity. The rest of the macropore film serves as a drainage ensuring the strength of RO-membranes.

The effective salt-retention by RO-membranes is determined by two factors: a lowered (as compared to values in a free solution) ion concentration in the selective layer and a higher (than for migrating ions) motion rate of the solvent (water) molecules through this layer [263]. The decrease in ion concentration may first of all be caused by the effect of a "nondissolving volume", i.e., by a decrease of the dissolving capacity of water in membrane submicropores due to its structurization. It is usually considered that a change in water structure (the ordering and decrease connected with it in effective dielectric penetrability) occurs in this (< 1 nm) pre-surface layer close to the hydrophilic surface and this determines the change of the standard chemical ion potentials in these layers [264].

The properties of the pre-boundary water layer have been studied by different physico-chemical methods—NMR, IR and X-ray spectroscopy, calculating the disjoining pressure and viscosity, and investigating thermal osmosis and capillary osmosis in thin pores [263,265–267]. The principal thermodynamic characteristic of thin films and layers is the disjoining pressure, the value of which is defined as the difference between the pressure acting upon the surface layer and the pressure in the liquid volume. In some approximations (for details see [265]) the disjoining pressure may be presented as a sum of three independent components: the dispersional (molecular), electrostatic and structural. The first two components may quite reliably be calculated theoretically. As to calculating the structural component, there are no reliable methods at present, and quantitative conclusions on structure and extent of preboundary layers are so far difficult to arrive at. Nevertheless, theoretical works in this direction (see [268, 269]) proving the existence of structural forces spreading to a distance of about 10–20 molecular radii without a doubt have heuristic value.

The structural component of the disjoining pressure was experimentally found during direct measurements of the interaction force between quartz and mica plates. In particular, it has been shown [270,271] that the repulsion force between cylinders covered with mica films starts exceeding the sum of electrostatic and molecular forces if the distance between cylinders is smaller than (4.0–7.5) nm. And the laws of the excess-force change conform to the experimentally established ones in the study of thin film stability. Papers [272,273] also provide analogous conclusions on the effects of liquid layer structurization in the vicinity of hydrophilic surfaces (the zone up to 16 nm with high water viscosity near the surface of quartz capillary was under investigation). Recently, investigations have shown that the structurization effects of liquid layers are observed not only near hydrophilic surfaces but also near hydrophobic ones [261, p. 369]. Indeed, the dissolution of an apolar substance in water causes a structure reorganization of the solvent (it becomes more ordered than the initial one) and a decrease in entropy.

Effects of water structurization in hydrophobic or weakly hydrophilic submicropores apparently play a noticeable role in the processes of reverse osmotic demineralization of the salt solutions. Cellulose acetate is a moderately hydrophilic material, its ion-exchange capacity is usually 10^{-2} mole eq/l, i.e., by two orders smaller than in ion-exchange membranes. The interaction of water molecules with the polymer matrix is much weaker than the interaction between water molecules as has been shown in the investigation of the water state in cellulose acetate at low moisture content using methods of IR spectroscopy [274] and Raman light scattering [275]. Therefore, in amorphous regions of cellulose acetate with low moisture content (less than 40%), the creation of water molecule dimers and trimers with particles filling the submicropores is preferable. At a higher moisture content the aggregation of water molecules in submicropores thermodynamically more advantageous including dimer and trimer transitions from hydrophobic regions into submicropores and the formation of a water submicrophase with characteristics close to that of water in large volume [275]. The action of "forcing through" pressures (about 5–10 MPa) in the RO-membrane active layer probably forms built-up submicrochannels with characteristic size $R \lesssim 1$ nm by means of which the directed transfer of water molecules and partly salt ions is achieved. (Salt retention coefficient r may reach 0.98). Unfortunately at present there are no direct data on the state of water and ions in the RO-membrane selective layer and on the submicropore distribution over size under conditions of membrane functioning. However, it is natural to assume that if the characteristic sizes of the channels are $R \lesssim 1$ nm, the enclosed water molecules in them appear structurizated due to noticeable hydrophoby and limited wetting ability of the matrixes. This must decrease their capacity to hydrate ions and lead to a considerable decrease of the ion distribution coefficient in the active layer phase.

Recent years have witnessed the creation of reverse osmotic membranes of a new generation which permit conducting natural (salty sea) water demineralization on a wide scale. The best functional indices are inherent in composition membranes obtained on microporous substrates from polysulphone during interphase polycondensation of aromatic diamine (e.g., m-phenylendiamine) with chlorineanhydrides of tricarbone acids (e.g., the tri-mezine or 1.3.5-cyclohexane-tri-carbone acid [276–278]). Such coatings of aromatic polyamide are characterized by high chemical, mechanical and thermal stability, which is accounted for by the rigid three-dimensional structure of the polymer molecule that resembles the structure of the disordered ceolyte. This also ensures a high water permeability ($\sim 1\ m^3/m^2$ day) with salt retention up to 99.5% at anomalously low for the reverse osmotic working pressures 2.0–3.0 MPa (such productivity was achieved at pressures 5.0–6.0 MPa on earlier generation membranes).

However, these membranes are characterized by insufficient resistance to sterilizing agents, including chlorine, which imposes certain limitations on the application of the membranes for water preparation and for processing industrial sinks. It is possible to enhance the resistance (stability) to oxidizers for polymer materials not containing NH-groups, in particular, polypiperasineamides. The corresponding reverse osmotic membranes (worked out by the firm "Separem") with a selectivity over NaCl about 95% and productivity ~ 1.0 m^3/m^2 day (working pressure about ~ 4 MPa) are stable in salt solutions at chlorine concentration 50.

If in the demineralization process the water percolates through a rigid system of submicropores (cavities) and channels of the reverse osmotic membrane selective layer, the salt retention is determined by both the size of the channel connecting hydrated cavities and a random force field created by all polymer fragments in the channel region.

To calculate the selective layer hydrodynamic resistance, one can use approaches based on the theory of liquid percolation in a random labyrinth which were earlier developed for problems of nonwetting liquid transfer (e.g., mercury) in porous materials [279]. According to [279] it is advisable to represent pore distribution in space as a three-dimensional lattice, the bonds (rits) of which correspond to pore channels and the points—to their intersections, i.e., cavities. Such an approach reduces the problem of mutual liquid and gas phase distribution (the cavities and channels respectively occupied and unoccupied by water molecules) to a classical problem of percolation theory—the problem of percolation over bonds.

Let us consider the process of a nonwetting liquid pressing into a pore system [279]. At an excess pressure 0, the nonwetting liquid can be forced through pore channels with an effective size exceeding some value R_+ that depends on P and the surface force field in the channel. We shall call such pore channels with $R > R_+$ supercritical. The part P of supercritical pore channels is determined by a function $\varphi(R)$ of channel distribution over size:

$$P = \int_{R_+}^{R_{max}} \varphi(R)dR \tag{5.34}$$

When considering the liquid phase in macroscopic approximations, it is convenient to introduce the notions of surface tension coefficient σ_s on the interface boundary liquid-vapor, and wetting edge angle θ. In the case of a cylindric pore $R = 2\sigma_s \cos\theta/P$.

At excess pressure P, the nonwetting liquid penetrates only those pores which are connected with the outer surface of the sample by a sequence of supercritical pore channels. It is known that at $P < P_c$ (where P_c is the lattice percolation threshold), the probability that a continuous sequence of supercritical channels exists decreases exponentially with an increase in the length of such a sequence. Then the nonwetting liquid penetrates only into the supercritical pores located in the immediate vicinity of the outer surface. Only when the excess pressure increases so that part P of the supercritical pore channels exceeds percolation threshold P_c, the nonwetting liquid penetrates into the sample volume. The corresponding pressure is called breakdown. Its value P_c is determined from the relation

$$P_c = \int_{R_+(P_c)}^{R_{max}} (R)\,dR \qquad (5.35)$$

The phenomenon of a breakdown (the penetration of the nonwetting liquid into the sample volume) at $P > P_c$ is accounted for by the formation of an infinite connected system of supercritical pores which is called an infinite cluster (IC). An IC includes cavities (points) and supercritical channels (ribs, or bonds), the cavities being connected by a continuous sequence of supercritical channels with the outer surface. For large-size samples containing a macroscopically large number of elementary pores, one can ignore edge effects and consider the IC to be uniformly distributed throughout the sample volume and the saturation (characterizes the relative volume of the liquid phase in the total volume of the pore and channel system) of the sample of finite sizes to be equal to the saturation of a semi-infinite medium.

We shall assume that different size pore channels are distributed randomly. Then the distribution of supercritical bonds in the lattice will also be random: the probaiblity that an arbitrarily chosen bond is supercritical equals the relative part of supercritical pore channels. The percolation theory considers the following IC characteristics. The connectivity function $Q_c(p)$— the relative part of bonds included in the infinite cluster among all the bonds in the lattice; percolation probability $Q_p(p)$—the relative part of points appearing in the IC among all the points in the lattice; relative conductivity $\sigma(p)$—the ratio of the conductivity of the lattice in which conductors constitute part p and insulators—part $(1 - p)$ to the conductivity of the lattice in which all bonds are conductors. The characteristics of the corresponding phase distribution are expressed through these functions. In this example the nonwetting liquid occupies the pores of the IC formed by supercritical bonds, and saturation λ is expressed by the connectivity function and percolation probability

$$\lambda = \nu_p A_p(p) + (1 - \nu_p) Q_c(p) \qquad (5.35)$$

where ν_p is the part of the cavity volume in the total pore volume, and $1 - \nu_p$ is the volume part of pore channels.

The effective coefficients of liquid transfer over a system of channels are proportional to the IC conductivity

$$D(\lambda) = D(1)\sigma(p) \qquad (5.36)$$

where $D(\lambda)$ and $D(1)$ are the effective coefficients of transfer at saturation λ and at complete occupation of pore space respectively.

The connectivity function, percolation probability and relative conductivity for a simple cubic lattice have a threshold nonlinear character, while at $p < p_c$ are identically equal to zero. Such behavior is typical of all lattices irrespective of their topology. Basically, the percolation threshold is determined by the coordination number of lattice z. The approximate equality: $p_c \approx 1.5\ z$ is known. In the vicinity of the percolation threshold, the following asymptotic scaling invariant dependencies are satisfied:

$$Q_p \sim Q_c \sim (p - p_c)^{\beta} \tag{5.37}$$

$$\sigma \sim (p - p_c)^{t} \tag{5.38}$$

The critical indices β and t of percolation probability and conductivity [279] lie within the limits $0.35 < \beta < 0.47$; $1.5 < t < 2.0$. Far from the percolation threshold, the linear equation of self-consistent field theory is valid for conductivity $\sigma(p)$

$$\sigma(p) = \left(p - \frac{2}{z}\right)\left(1 - \frac{2}{z}\right)^{-1} \tag{5.39}$$

Percolation probability $Q_p(p)$ and connectivity function $Q_c(p)$ can be approximated by a power-like dependence; at $p > 2p_c$ they differ slightly from the asymptotic dependencies:

$$Q_p(p) \longrightarrow 1, \qquad Q_c(p) \longrightarrow p \tag{5.40}$$

According to the scaling invariance hypothesis the geometry of a percolation system, including the IC structure, is determined only by the characteristic size: correlation length L. Far from the percolation threshold at $(p - p_c)/p_c \approx 1$, the correlation length is of the same order as size $\bar{R}_1$ of one structural element. At $p \to p_c$ the correlation length increases limitlessly according to the power law:

$$L \sim (p - p_c)^{-1+\delta} \tag{5.41}$$

The critical index of correlation length for three-dimensional lattices is $(1 - \delta) \simeq 0.9$ [279]. At distances $r \gg L$ the infinite cluster is homogeneous, while at $r \ll L$ one may observe the correlation in the distribution of elements belonging to the IC. In the vicinity of the percolation threshold when $L \gg \bar{R}_1$ region $\bar{R}_1 \ll r \ll L$ exists in which the IC exhibits so-called fractal properties, i.e., it is characterized by a nontraditional dependence of its volume (or surface) on the geometric scale. For example, the number of elements for the IC located at distance $r' < r$ from the selected element increases proportionally to r^{d_f} where the so-called fractal dimensionality d_f is not equal to space dimensionality d:

$$d_f = d - \beta/(1 - \delta) \tag{5.42}$$

(in the three-dimensional case $d_f = 2.5$). Ideas about the fractal properties of the IC prove useful when investigating the geometric specificities of reciprocal phase distribution in pore space.

For calculating the effective transfer coefficients in partially saturated porous media, the dependence of relative conductivity σ on the relative part of elements contained in the IC is quite interesting. It follows from scaling relations (5.37) and (5.38) that

$$\sigma \sim Q_p^{t/B} \sim Q_c^{t/B} \tag{5.43}$$

Equation (5.35) implies that in the scaling region effective coefficients of transfer through the nonwetting phase are connected by a power-like dependence with the saturation, the index of power being equal to the ratio between the critical indices of relative conductivity and percolation probability

$$D(\lambda) \sim \lambda^{t/B} \tag{5.44}$$

The authors of [279] also derived an interpolation expression for $D(\lambda)$:

$$D(\lambda) = D(1)\,\frac{2{,}88\cdot\lambda^{3,48}}{(1 + 1{,}31\nu_p)^{3,48}} \qquad \text{at} \qquad \lambda < 0{,}433(1 + 1{,}31\nu_p) \tag{5.45}$$

Equations (5.44–5.45) provide a basis for analyzing the regularities of penetration of the nonwetting liquid into porous systems depending on the applied pressure. However, the corresponding analysis of the water permeability of reverse osmotic membranes requires data on the parameters of the selective layer structure, including data on the function of pore distribution over size, and on the parameters of the surface force field. Unfortunately, such data is practically unavailable at the present time. Corresponding investigations will permit adequate estimates of the potential possibilities of the reverse osmotic method for separating various mixtures, including water-organic and organic liquid ones. At present dependencies of types (5.44–5.45) have been established for processes of gas and electrolyte transfer in porous nickel electrodes.

Ultrathin surface layers of submicroscopic (∼ 10 nm) thickness from organic and inorganic (e.g., polyphosphasenes) materials formed on ultra- or micro-filtration substrates may serve as barrier structures with fractal properties when separating electrolyte mixtures, gas mixtures in the conditions intensive separation regimes, and in electrocatalysis processes. In this case not only the corresponding permeability coefficients, but also the parameters of separation selectivity, depend on the thickness of the subthin layer. Methods for preparing such structures may be connected with applying the Langmuir-Blodget methods of forming monomolecular films and mixed structures with a high degree of molecule ordering, and the electrochemical, plasmochemical and photochemical methods of applying subthin layers [280]. This actually means a qualitatively new level of membrane technology, since the membrane permeability over the desired components increases by more than an order, and consequently the specific consumption of materials in the membrane process considerably decreases. Creation of such barrier structures and subthin layers requires a superpure technology (with respect to materials, and concentration of aerosols).

5.3. MEMBRANE GAS SEPARATION

The solution of many problems in chemical technology, petrochemistry, the gas industry, environmental protection, the food industry, and in various other branches of the economy is connected with the separation, purification or concentration of gas mixtures. The method most widely used at present for gas separation is the cryogenic one. However, in recent years interest towards membrane methods of separation and cleaning of mixtures has grown considerably. This is, first of all, connected with the creation of new membrane materials possessing high separation properties (penetrability, selectivity [215]). The wide use of membrane methods of gas separation was promoted by the new technology for obtaining highly productive asymmetric membranes in which the separation capacity is inherent only in a thin (< μm) selective layer on the porous drainage substrate from the same material [281]. Using hollow asymmetric fibers as membranes made possible the design of compact gas separators. At present, such apparatus are being successfully employed for solving various problems, in particular for hydrogen extraction from circular gas mixtures in the production of ammonia and methanol [281,282]. The working pressures in the corresponding gas-separation installations reach (15–30) MPa.

Membrane apparatuses for air oxygenation with oxygen content up to 30–45% (vol.) are widely found (in microbiology, in the technology for biological purification of sewage, in the fish industry, and in metallurgical production) [276]. Membrane gas separation installations require little energy and they are economical and simple to use.

The basis for membrane gas separation is formed on the notions of the diffusion transfer of gas mixture molecules through nonporous, usually polymer, membranes. Diffusion coefficients D_i and distribution constants δ_i (Henry constants) of different i components of the gas mixture are different, which upon an appropriate organization of membrane separation provides for extracting the required components from the system. We shall not discuss general questions on the diffusion transfer of gas molecules through polymer membranes, neither shall we dwell upon problems of optimizing membrane processes (see for example monographs [214–217]). Here we attach special attention to an analysis of new tendencies in membrane gas separation: gas separation under conditions of high pressure and a change in the parameters of polymer membranes (gas permeability, selectivity) at surface and volume polymer material modifications.

The influence of interphase boundaries on transmembrane gas transfer processes was first discovered in a hydrogen-metal membrane system. It appeared possible to control the H_2 permeability through metal partitions within large limits [283] by changing the state of the metal-gas interface. It is then possible for the hydrogen flow falling to the membrane and passing through it to be values of the same order (super-permeability) [284]. Strictly speaking, the effects of the influence of the interface on gas transmembrane transfer are specific not only for the hydrogen-metal system. They can also be observed in other gas-solid phase membrane systems. It will be shown below that conditions for gas molecule transport in the interphase boundary region can to considerable extent determine the gas permeability of polymer membranes.

5.3.1. Gas Permeability of Membranes Based on Siloxane-Containing Block Copolymers

Paper [285] experimentally investigated transfer processes of simple gases (O_2, N_2, CO_2) through membranes based on polyarylensulphonoxidepoly-dimethylsiloxane block copolymer (PSS) [286] with the following characteristics: rigid block mass $4.6 \cdot 10^3$; mass of flexible block $4 \cdot 10^3$; molecular mass of the block copolymer $12 \cdot 10^4$; vitrification temperature of the rigid phase 448 K, of the flexible phase—150 K. The choice of the initial material was determined by the large values of gas permeability for siloxane-containing block copolymers and their wide application in membrane technology for gas separation. To change the conditions of gas molecule transport in the region of interphase boundaries, gas—polymer PPS-membranes were modified by fluorine-containing allyloxysilane (AS)—general formula $C_{16}H_{16}F_{12}O_4Si$.

The investigated membranes were moulded from a PSS solution in chloroform. And both the original and modified membranes were obtained in the same technological cycle. PSS-membrane modification was obtained using two methods. The first consists of applying the AS solution in chloroform on the surface of the original membrane. The obtained sample was dried in an oxygen atmosphere at $T = (293 \div 385)$ K. According to the second method, the AS is added into the initial solution of the polymer and then the membrane was moulded at 293 K.

Upon introducing AS molecules they "sweat out" on the interphase boundary gas—polymer, which can be registed by changing the wetting angle of the boundary with a liquid (the angle of wetting by water of the initial and modified membrane surface is $\theta_0 = 45$ and $\theta_M = 100°$). AS molecules appearing on the surface can be polymerized (due to double C=C bonds) in an oxygen atmosphere at 293 385 K and thus fixed on the interphase boundary. Note that the complete filling* of the surface by AS molecules requires a relatively small amount (∿ 1% from the total mass) of the modificator.

When applying an AS solution on the surface of the original polymer membrane, the modifying layer is formed in the process of sample drying. Note that the boundaries of such membranes are not identical (membranes are modified on one side).

The gas permeability coefficients of the modified membranes are about twice as high (Table 5.2) as those of the original ones†, while gas diffusion coefficients remain practically unchanged. It should be noted that the difference in the drying regimes and the change in modificator concentration within the limits of $(2 \div 6)$ % from membrane mass does not influence the measured values. This testifies to the fact that the AS modifies the interphase boundary polymer—gas and not the mebrane matrix. To check this conclusion, a series of membranes modified by introducing

Table 5.2. The values of gas permeability Π and gas diffusion coefficients D(293 K) in PSS-membranes with a modified surface

Membrane	$\Pi \cdot 10^5$, mole m/(m^{2}s P_a)			$D \cdot 10^{10}$, m^2/s		
	O_2	N_2	CO_2	O_2	N_2	CO_2
Original, l = 104 μm	30	14	240	1.9	1.8	1.6
Modified, l = 110 μm	61	33	345	1.8	1.8	1.8

Table 5.3. The values of gas permeability Π and diffusion coefficients D(293 K) in PSS-membranes upon introducing the modificator into the polymer solution

Membrane	$\Pi \cdot 10^{15}$, mole m/(m^{2}s P_a)			$D \cdot 10^{10}$ m^2/s			θ deg
	O_2	N_2	CO_2	O_2	N_2	CO_2	
Original, l = 53 μm	31	17	260	1.9	1.8	1.6	45
Modified* l = 53 μm, *t* = 2 months	31	17	260	1.9	1.8	1.6	45
the same, t = 12 months	63	33	280	1.8	1.7	1.8	90

*modified by introducing AS

AS into the original polymer solution were studied. Such a modification did not reflect upon the membrane parameters for a long time (2–3 months months): the coefficients of gas diffusion and permeability, as well as the the wetting angle of the surfaces of initial and modified membranes remained the same. However, after 10–12 months gas permeability coefficients increased approximately twice, the wetting angle of both surfaces changed, while the diffusion coefficient practically remained constant (Table 5.3). Such behavior of values Π, D and θ shows that by sweating

*Complete filling here means a situation when the further introduction of the modificator does not practically change either the wetting angle or the gas permeability coefficient.

†Measurements of gas permeability and gas diffusion coefficients were carried out according to the methods of [287]. According to [287] a membrane is considered a uniform partition if it has negligibly-small diffusion entrance and exit resistances. Gas pressure jump on the membrane was P = 0.1 MPa.

out on the surface with time, the AS modifies the interphase boundary gas—polymer without appreciably changing the volume characteristics of the membrane matrix.

The time required for stabilizing the state of the interphase boundaries (AS "sweating out") is defined by two processes: the transport of AS molecules from the polymer to the boundary, as well as by the volatility of these molecules from the interface.

In order to reveal the specificity for the influence of surface effects on membrane gas permeability, it is necessary to solve, in the general form, the problem of gas molecule transfer through a semipermeable partition with nonidentical boundaries.

5.3.2. Model Description of Transmembrane Gas Transfer

Let axis x of the rectangular coordinate system (Figure 5.7) be perpendicular to the surface of a membrane occupying space $0 \leqslant x \leqslant l$. The gas flow j_M through the memebrane we will represent in the form of the known [1] relation

$$j_M = -\frac{1}{KT} L(x) \frac{\partial \mu}{\partial x} \tag{5.46}$$

where $\mu = KT \ln c + \psi(x, T)$ is the chemical potential of the dissolved gas; $c(x, t)$ is the concentration of the molecules (atoms) of this gas in point x at time t; L is the kinetic coefficient; $\psi(x, T)$ is the function on the coordinate and temperature.

Introducing diffusion coefficient $D = L/c$ we derive the expression for the flow in the form

$$j_M = -D\frac{\partial c}{\partial x} - D\frac{c}{KT}\frac{\partial \psi}{\partial x} \tag{5.47}$$

Taking into account the continuity equation, the relation (5.47) gives

$$\frac{\partial c}{\partial t} = \frac{\partial}{\partial x}\left(D\frac{\partial c}{\partial x} + D\frac{c}{KT}\frac{\partial \psi}{\partial x}\right) \tag{5.48}$$

Taking into account the term with ψ_x' in the relation for flow (5.47) may prove essential when considering transfer through spatially nonuniform systems, in particular, when analyzing gas molecule transfer through membranes characterized by dependencies $D = D(c)$ or $D = D(x)$.

To make the problem under consideration more definite it is necessary to supplement (5.48) with the corresponding boundary and initial conditions The boundary conditions for diffusion problems of mass transfer through the interface are in the most general form formulated in Chapter 2. Let us represent the state of a gas molecule on the interphase boundary as a complex of two states: I—boundary proper (characterized by surface concentration n_0 or n_l) and contiguous state II (characterized by volume concentration c) which initiate the process of random walks (see Figure 5.7). The corresponding equations of material balance (boundary conditions) take the form:

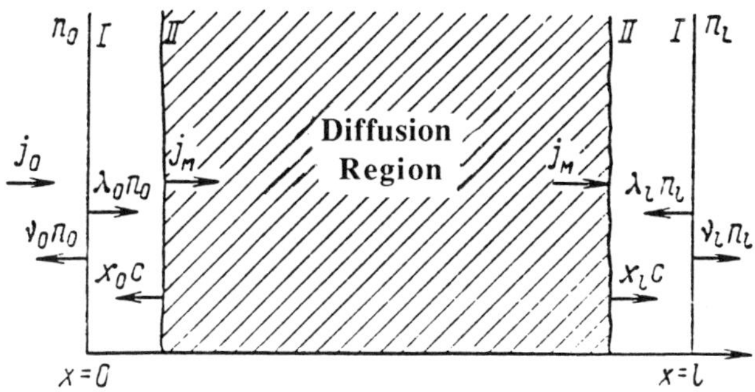

Fig. 5.7. Scheme of gas transfer through semipermeable partitions.

$$\left.\begin{aligned} \frac{dn_0}{dt} &= j_0 + \chi_0 c - \lambda_0 n_0 - \nu_0 n_0 \\ \lambda_0 n_0 &= \chi_0 c + j_M \end{aligned}\right\} \quad x = 0 \tag{5.49}$$

$$\left.\begin{aligned} \frac{dn_l}{dt} &= \chi_l c - \lambda_l n_l - \nu_l n_l + j_l \\ j_M &= \chi_l c + \lambda_l n_l \end{aligned}\right\} \quad x = l \tag{5.50}$$

where $j_{0(l)}$ is the molecule flow from the gas phase onto surface $x = 0(l)$; $\lambda_{0(l)}$, $\chi_{0(l)}$ and $\nu_{0(l)}$ are the rate constants of the corresponding transitions on these surfaces.

Let us consider the stationary process of gas transfer through the membrane. In this case, we obtain the boundary problem from (5.47–5.50)

$$\frac{d}{dx}\left(D\,\frac{dc}{dx} + D\,\frac{c}{KT}\,\psi_x'\right) = 0$$

$$j_0 + D_0\,\frac{dc}{dx} = [x_0 - D_0\psi_{x=l}'(KT)^{-1}]c \qquad \text{at} \qquad x = 0 \tag{5.51}$$

$$j_l - D_l\,\frac{dc}{dx} = [x_l + D_l\psi_{x=l}'(KT)^{-1}]c \qquad \text{at} \qquad x = l$$

where

$$x_{0(R)} = \alpha_{0(l)}\chi_{0(l)}, \qquad \alpha_{0(l)} = \nu_{0(l)}\lambda_{0(l)}$$

$$\psi'_{x=0(l)} = \left.\frac{\partial\psi}{\partial x}\right|_{x=0(l)},$$

$$D_0 = (1+\alpha_0)D(0), \qquad D_l = (1+\alpha_l)D(l)$$

Note that in the general case with considerable deviations from equilibrium (see below) of the gas—polymer system, the rate constants introduced depend on flow j_M and $j_{0(l)}$, i.e.,

$$\lambda_{0(l)} = \lambda_{0(l)}(j_M, j_{0(l)}), \qquad \chi_{0(l)} = \chi_{0(l)}(j_M, j_{0(l)})$$

In the equilibrium case, when $j_M = 0$, values $\lambda^*_{0(l)} \equiv \lambda_{0(l)}(0, j_{0(l)})$, $\chi^*_{(0,l)} = \chi_{0(l)}(0, j_{0(l)})$, $\nu^*_{0(l)} = \nu_{0(l)}(0, j_{0(l)})$ and solubility coefficient σ^* (Henry's constant) are interconnected by the relation

$$\sigma^* = \frac{1}{N_A\sqrt{2\pi mKT}}\ \frac{\lambda^*_0}{\nu^*_0\chi^*_0}\ \frac{1}{l}\int_0^l \exp\left[-w(x)\right]dx \tag{5.52}$$

where N_A is the Avogadro number; m is mass of a gas molecule; $w(x) = (KT)^{-1}[\psi(x) - \psi(0)]$.

At this $x^*_0/x^*_l = \exp(w_l)$, where $x^*_{0(l)} \equiv x_{0(l)}(0, j_{0(l)})$, which reflects the Boltzmann distribution of gas molecules in an equilibrium system.

Solving system (5.51) one can find the equation of gas flow j_M through half-permeable partition

$$j_M = \frac{j_0}{x_0(j_M)}\left[\frac{1+\alpha_0(j_M)}{x_0(j_M)} + F_l + \frac{1+\alpha_l(j_M)}{x_1(j_M)}\exp w_l\right] \tag{5.53}$$

$$F_l = \int_0^l D^{-1}(x)\exp\left[w(x)\right]dx, \qquad w_l \equiv w(l)$$

When deriving (5.53) it was considered that $j_l = 0$ (vacuum is on one side of the partition).

It is necessary to obtain the final expression for flow j_M setting the explicit form of functions $x_{0(l)}(j_M)$ and $\alpha_{0(l)}(j_M)$. It follows from (5.53) that the dependence of values $x_{0(l)}$ and $\alpha_{0(l)}$ on flow j_M yields the dependence of permeability coefficient Π on gas pressure jump on the membrane (the case when $x_{0(l)} = x_{0(l)}(j_M/j_{0(l)})$ and $\alpha_{0(l)} = \alpha_{0(l)}(j_M/j_{0(l)}(j_M/j_{0(l)})$ could be the only exception).

Figure 5.8 presents the experimental values for the permeability coefficients of the initial and modified membranes over O_2 and N_2 at different pressure jumps. It can be seen that gas permeability coefficients are practically independent of ΔP within a wide pressure range. However, in the region of small values $\Delta P(\approx 2KPa)$ the gas permeability coefficients of membranes based on siloxane-containing block copolymers increase considerably (by several times) with the ΔP increase reaching the limit value at $\Delta P = 4KPa$ [288].

Such behavior of Π points out the fact that for siloxane-containing membranes nonequilibrium effects with gas transfer through the interphase boundary are revealed already at pressure $\sim 2KPa$, i.e., in the atmospheric pressure region the ratio Π/D appears to be a nonequilibrium value. Dependencies $\Pi(\Delta P)$ found in [288] may be qualitatively described on the basis of relation (5.53) taken that $x_{0(l)}{}^{-1}(1 + \alpha_{0(l)}) \ll F_l$ and $x_0 = x^*{}_0(\gamma j_M + j_\infty)/(j_M + j_\infty)$ (where j_∞ is the characteristic value of flow j_M beginning from which parameter x_0, after reaching maximum value $\gamma x^*{}_0$, becomes independent of the gas flow through the membrane).

Thus, taking into account the above-mentioned specificity of the behavior of experimental dependencies $\Pi(\Delta P)$, it can be considered that within the pressure range $(0.1 \div 15)$ MPa analyzed values $x_{0(l)}$ and $\alpha_{0(l)}$ are independent of flow j_M.

The requirement on the equality of gas flows through the membrane at different orientations with respect to regions of high and low pressures (the valve effect is absent in the considered systems) gives the relation

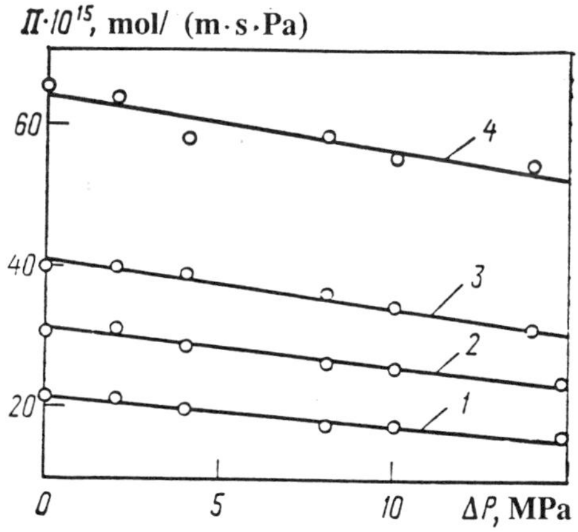

Fig. 5.8. Dependence of coefficients of nitrogen (1,2) and oxygen (3,4) permeability through the initial (1,3) and modified (2,4) PSS membrane on pressure jump $\Delta P(T = 293$ K).

$$x_0/x_l = \exp(w_l) \tag{5.54}$$

Then it follows from (5.53) and (5.54) that

$$j_M = j_0/[x_0F_l + (2 + \alpha_0 + \alpha_l)] \tag{5.55}$$

In further discussion, we assume that diffusion coefficient D is independent of coordinate x, since neither the modification nor the membrane thickness l effect the experimental values of D (see Figures 5.8 and 5.9).

From (5.55) we obtain the expression for value Π

$$\Pi = \frac{1}{N_A\sqrt{2\pi mKT}} \frac{Dl}{x_0H_l + (2 + \alpha_0 + \alpha_l)D} \tag{5.56}$$

$$H_l = \int_0^l \exp[w(x)]dx = l\exp[w(\xi)]$$

At low surface resistances when $x_0H_l \gg (2 + \alpha_0 + \alpha_l)D$ relation (5.81) can be presented in the form

$$\Pi = D\sigma_{eff} \tag{5.57}$$

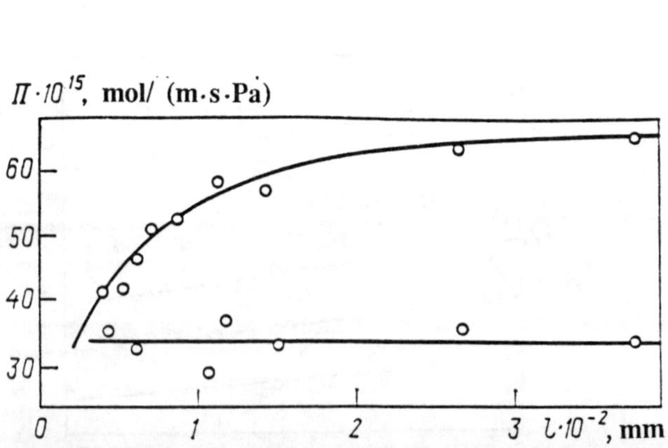

Fig. 5.9. Dependence of the oxygen permeability coefficients through PSS membranes on thickness 1 at 293 K and ΔP = 0.1 MPa. 1—is the original membrane; 2—is the modified one; continuous lines show the calculation according to expression (5.56) [291].

Table 5.4. The value of gas permeability coefficients for the initial and modified membranes

Membrane	O_2		N_2		CO_2	
	$\sigma_{eff}10^{15}$	$\alpha_0 + \alpha_l$	$\sigma_{eff}10^{15}$	$\alpha_0 + \alpha_l$	$\sigma_{eff}10^{15}$	$\alpha_0 + \alpha_l$
Original	19	$4\cdot10^6$	12	$7\cdot10^6$	225	$3\cdot10^5$
Modified	44	$1.4\cdot10^6$	16	$2.6\cdot10^6$	249	$1.1\cdot10^6$

where $\sigma_{eff} = (N_A\sqrt{2\pi mKT}\ x_0)^{-1}l^{-w(\xi)}$ is the effective coefficient of gas solubility in the membrane.

Note that the introduced value σ_{eff} differs in the general case from the true Henry constant σ^*, which connects the equilibrium concentration of gas molecules in the membrane with the pressure of this gas (5.52). The difference mentioned between σ^* and σ_{eff} may be the source of discordance when defining the gas permeability coefficient by the direct method and on the basis of indirect measurements, carried out using the gravimetric method.

The obtained expressions (5.56–5.57) provide a basis for understanding all the given data on the changes in gas permeability acquired by modifying the membrane surface. The analysis of these expressions shows that no surface modification can change value Π of a sufficiently thick $\{l \gg (2 + \alpha_0 + \alpha_l)Dx_0^{-1}\exp[-w(\xi)]\}$ membrane, if the mass transfer passes in a quasi-equilibrium regime $(D\sigma^*\Delta P/l \ll j_\infty)$. Surface effects may be made effective if we achieve a regime of nonequilibrium transport in the region of interphase boundaries when $j_\infty < D_{eff}\Delta P/l$. Then the gas permeability coefficient Π of the thick membrane changes. Thus it follows that proceeding to materials with a high gas permeability (siloxane-containing block copolymers) promotes the realization of a nonequilibrium transfer regime and the efficiency of surface modifications. In the case of thin membranes $\{l \ll (2 + \alpha_0 + \alpha_l)Dx_0^{-1}\exp[-w(\xi)]\}$ the surface treatment changing parameter $\alpha_{0(l)}$ may also lead to a change in value Π. Such a situation is apparently relaized with adsorption modifications of metal membranes [302].

It follows from (5.56) that the gas permeability coefficient Π of the membrane depends on its thickness.

Figure 5.9 compares this dependence to the corresponding experimental results on the oxygen permeability through PSS-membranes. Such a comparison provides for evaluating values $(\alpha_0 + \alpha_l)$ appearing in expressions (5.56–5.57). Estimates of these values obtained using the least-square technique are shown in Table 5.4.

Note that the values found of σ_{eff} have the same order as the corresponding values of the ratio Π/D for membranes made from siloxane-containing block copolymers with a close relationship between the mass of the flexible and rigid blocks [290]. Thus by changing the conditions for gas-molecule transport in the region of interphase boundaries the gas-polymer may purposefully affect the gas permeability of polymer membranes.

REFERENCES

1. de Groot, S. R., Mazur P., *Nonequilibrium Thermodyanmics.* North-Holland Publishing Company, Amsterdam. 1962.
2. Crank, J., *The Mathematics of Diffusion.* Oxford, Clarendon Press, 1975, 414 p.
3. Frank-Kamenetskii, D. A., *Diffuziya i teploperedacha v khimicheskoi kinetike* (in Russian) (Diffusion and Heat Transfer in Chemical Kinetics). Moscow, USSR, Nauka Publishing House, 1987, 491 p.
4. Tunitskii, N. N., *Diffuziya i sluchaynie protsessy* (in Russian) (Diffusion and Random Processes). Novosibirsk, Nauka, 1970, 116 p.
5. Tunitskii, N. N., Kaminskii, V. A., Timashev, S. F., *Metody fiziko-khimicheskoi kinetiki* (in Russian) (Methods of Physical-Chemical Kinetics). Moscow, Khimiya, 1972, 197 p.
6. Noyes, R. M., *In*: *Progress in Reaction Kinetics,* 1961, vol. I, pp. 129–160.
7. Kapitanskii, V. R., et al, *ZhTF* (in Russian) (JTP), 1976, vol. 46, no. 5, pp. 1030–1041.
8. Belyi, A. A., Ovchinnikov, A. A., Timashev, S. F., *Zhur. Fizich. Khim.* (in Russian) (J. Phys. Chem.), 1979, vol. 53, no. 4, pp. 948–952.
9. Zwanzig, R., *Phys. Rev.,* 1961, vol. 124, no. 4, pp. 983–992.
10. Bashkirov, A. G., Zubarev, D. N., *Teor. i mat. fizika* (in Russian) (Theor. and Mat. Physics), 1969, vol. I, no. 3, pp. 407–419.
11. Pines, D., Nozieres, P., *The Theory of Quantum Liquids, Vol. I: Normal Fermi Liquids,* W. A. Benjamin, Inc., N.Y.-Amsterdam, 1966.
12. Zeldovich, Ya. B., Ovchinnikov, A. A., *ZETF* (in Russian) (JETP), 1978, vol. 74, no. 5, pp. 1588–1598.
13. Zeldovich, Ya. B., Ovchinnikov, A. A., *Pis'ma v ZhETF* (in Russian) (JETP Letters), 1977, vol. 26, no. 8, pp. 588–591.
14. Emanuel, N. M., Knorre, D. G., *Kurs khimicheskoi kinetiki* (in Russian) (Course on Chemical Kinetics). Moscow, Vysshaya Shkola, 1974, 400 p.
15. Barret, P., *Cinetique Heterogene Gauthier.* Villars, 1973, Paris.
16. Wilemski, G., Fikman, M., *J. Chem. Phys.,* 1973, vol. 58, no. 9, pp. 4009–4019.
17. Kupperman, A., *Nucleonics*, 1961, vol. 19, no. 10, pp. 38–42.
18. Pshezhetskii, S. Ya., *Mekhanizm i kinetika radiatsionno-khimicheskikh reaktsii* (in Russian). (The Mechanism and Kinetics of Radiation Chemical Reactions), Moscow, Khimiya, 1968, 368 p.
19. Chandrasekhar, S., *Reviews of Modern Phys.,* 1943, vol. 15, no. I, pp. 1–89.
20. Eyring, H., Lin, S. H., Lin, S. M., *Basic Chemical Kinetics.* John Wiley and Sons, N.Y.-Chichester-Brisbane-Toronto, 1980.
21. Fehder, P. L., Emies, C. A., Futrelle, R. P., *J. Chem. Phys.,* 1972, vol. 54, no. II, pp. 4921–4934.

22. Wait, T. R., *Ibid*, 1958, vol. 28, no. I, pp. 103–106.
23. Wait, T. R., *Phys. Rev.*, 1957, vol. 107, no. 2, pp. 463–470.
24. Monchik, L., *J. Chem. Phys.*, 1956, vol. 24, no. 2, pp. 381–385.
25. Zeldovich, Ya. B., Myshkis, A. D., *Elementy matematicheskoi fiziki* (in Russian) (Elements of Mathematical Physics). Moscow, Nauka, 1973, 351 p.
26. Smolukhovskii, M., *V kn.: Koagulyatsiya kolloidov* (in Russian) (In: The Coagulation of Colloids). Moscow, United Scient. Techn. Publ. House, 1936, pp. 7–39.
27. Yguerabide, J., *J. Chem. Phys.*, 1964, vol. 40, no. 10, p. 3041–3052; 1967, vol. 47, no. 8, pp. 3049–3061.
28. Grassberger, P., Procaccia, I., *J. Chem. Phys.*, 1982, vol. 77, no. 12, pp. 6281–6284.
29. Collins, F. C., Kimball, G. E., *J. Colloid. Sci.*, 1949, vol. 4, pp. 452–437.
30. Denisov, E. T., *Kinetika gomogennykh khimicheskikh reaktsii* (in Russian) (The Kinetics of Homogeneous Chemical Reactions). Moscow, Vysshaya Shkola, 1978, 367 p.
31. Burlatsky, S. F., et al., *Chem. Phys. Lett.*, 1979, vol. 66, no. 3, pp. 565–569.
32. Perico, A., Battezzati, M., *J. Chem. Phys.*, 1981, vol. 75, no. 9, pp. 4430–4434.
33. Nemzek, T. L., Ware, W. R., *Ibid.*, 1975, vol. 62, no. 2, pp. 477–489.
34. Steavens, B., *J. Phys. Chem.*, 1981, vol. 85, no. 24, pp. 3442–3557.
35. Chuang, T., Hofman, G., Eisental, K., *Chem. Phys. Letters*, 1974. vol. 25, no. 2, pp. 201–205.
36. Walling, G., Lepley, A., *Int. J. Chem. Kinetics*, 1971, vol, 3, no. 2, pp. 97–104.
37. Buchachenko, Z. Z., *Uspekhi khim* (in Russian) (Adv. in Chem.), 1979, vol. 48, no. 10, pp. 1713–1746.
38. Berg, O. G., *Chem. Phys.*, 1978, vol. 31, no. I, pp. 47–57.
39. Nakvi, K. R., Mork, K. J., Waldenstrom, S., *J. Phys. Chem.*, 1980, vol. 84, no. II, pp. 1315–1319.
40. Leontovich, M. A., *Vvedenie v termodinamiku* (in Russian) (Introduction to Thermodynamics). Moscow, Nauka, 1983, p. 416.
41. Mozumder, A., *J. Chem. Phys.*, 1978, vol. 69, no. 4, pp. 1384–1391.
42. Szabo, A., Schulten, Z., *Ibid*, 1980, vol. 72, no. 8, pp. 4350–4357.
43. Sano, H., Tachiya, M., *Ibid.*, 1979, vol. 71, no. 3, pp. 1276–1282.
44. Sano, H., *Ibid.*, 1984, vol. 74, no. 2, pp. 1394–1400.
45. Blumen, A., Zumofen, G., *Ibid.*, 1981, vol. 75, no. 2, pp. 892–907.
46. Rytov, S. M., *Sluchainye Protsessy* (in Russian) (Random Processes). Moscow, Nauka, 1976, pt. I, p. 494.
47. Morse, P., Feshbach, H., *Methods of Theoretical Physics*, McGraw-Hill Book Company, Inc., N.Y.-Toronto-London, Part 1–8, 1953.
48. Sobolev, S. Z., *Uravneniya matematicheskoi fiziki* (in Russian), (Equations of Mathematical Physics). Moscow-Leningrad, OGIZ, 1947, p. 440.
49. Vrunov, P. A., Kaminskii, V. A., Timashev, S. F., *Teoret. i exper. kim* (in Russian) (Theor. and Exper. Chem.). 1974, vol. 10, no. 3, pp. 380–384.

50. Freeman, D. L., Doll, J. D., *J. Chem. Phys.*, 1983, vol. 78, no. 10, p. 6002–6009.
51. Doi, M., *J. Chem. Phys.*, 1975, vol. 9, no. 3, pp. 455–466; vol. II, no. 1, pp. 107–113; 115–121.
52. Wilemski, G., Fixman, M., *J. Chem. Phys.*, 1974, vol. 60, no. 3, pp. 866–890.
53. Berdinikov, V. M., Doktorov, A. B., *Teoret. i eksper. kim.* (in Russian) (Theoret, and Exper. Chem.), 1981, vol. 17, no. 3, pp. 318–326.
54. Doktorov, A. B., Yakobson, B. I., *Chem. Phys.*, 1981, vol. 60, no. 2, pp. 223–230.
55. Burshtein, A. I., Yakobson, B. I., *Int. J. Chem. Kinetics*, 1980, vol. 12, no. 4, pp. 261–270.
56. Solc, K., Stockmayer, W. H., *J. Chem. Phys.*, 1971, vol. 54, no. 7, pp. 2981–2988.
57. Kac, M., *Probability and Related Topics in Physical Sciences.*, Interscience Publishers, Ltd., N.Y.-London., Inc., 1957.
58. Shieder, W., *J. Phys. Chem.*, 1972, vol. 76, no. 3, pp. 349–360.
59. Khudyakov, I. V., Yakobson, B. I., *ZhOKh* (in Russian) (J. of Organic Chem.), 1984, vol. 54, No. 1, pp. 3–23.
60. Temkin, S. I., Yakobson, B. I., *J. Phys. Chem.*, 1984, vol. 88, no. 13, pp. 2679–2682.
61. Melvin-Hews, E. A., *Ravnovesie i kinetika reaksit v rastvorakh* (in Russian) (Equilibrium and Reaction Kinetics in Solutions) Moscow, Khimia, 1975, 470 p.
62. Wintle, H., *In*: *The Radiation Chemistry of Macromolecules*, Ed. M. Dole, vol. 1, Academic Press, N.Y.-London, 1972.
63. Vrunov, P. A., Timashev, S. F., *Teoret. i eksper. khim.* (in Russian) (Theor. and Exper. Chem.). 1978, vol, 14, no. 5, pp. 700–702.
64. Burlatsky, S. F., Ovchinnikov, A. A., *ZFK* (in Russian) (J. Phys. Chem.), 1980, vol. 54, no. 7, pp. 1741–1744.
65. Agekyan, T. A., *Teoriya veroyatnostei dlya astronomov i fizikov* (in Russian) (The Theory of Probability for Astronomers and Physicists). Moscow, Nauka, 1974, 264 p.
66. Davies, R. W., *Phys. Rev.*, 1954, vol. 93, no. 6, pp. 1169–1170.
67. Mozumder, A., *J. Chem. Phys.*, 1968, vol. 48, no. 4, pp. 1659–1665.
68. Jacobson, S. A., Ratner, M. A., Nizan, A., *Ibid.*, 1982, vol. 77, no. II, pp. 5752–5776.
69. Wilemski, G., *J. Statist. Phys.*, 1976, vol. 14, no. 2, pp. 153–169.
70. Benderskii, V. A., et al, *Pis'ma v thETF* (JETP Letters), 1980, vol. 32, no. 6, pp. 429–432.
71. Barelko, V. V., et al, *Khim. fizika* (in Russian) (Chem. Physics), 1983, no. 7, pp. 980–990.
72. Vlasov, A. A., *Statisticheskie funktsii raspredeleniya* (in Russian) (Statistical Functions of Distribution). Moscow, Nauka, 1966, 355 p.
73. Van Kampen, V. G., *Physica*, 1957, vol. 23, no. 8, pp. 707–816.
74. Sirovich, L., *The Physics of Fluids*, 1963, vol. 6, no. 2, pp. 218–230.
75. Monchick, L., *J. Chem. Phys.*, 1983, vol. 78, no. 4, pp. 1808–1814.

76. Weissman, M. B., Wave, B. R., *Ibid.*, 1978, vol. 68, no. II, pp. 5069–5076.
77. Haan, W., Zwanzig, R., *Ibid.*, 1978, vol. 68, no. 4, pp. 1879–1883.
78. Davydov, A. S., *Kvantovaya mekhanika* (in Russian) (Quantum Mechanics) Moscow, GIFML, 1983, 748 p.
79. Kittel, C., *Quantum Theory of Solids*, Wiley, N.Y.-London, 1963, 435 p.
80. Ovchinnikov, A. A., Erikhman, N. S., *ZhETF* (JETP), 1974, vol. 67, no. 4(10), pp. 1474–1483.
81. Kitahara, K., et al, *J. Chem. Phys.*, 1976, vol. 65, no. 7, pp. 2871–2882.
82. Klyatskin, V. I., *Stokhasticheskie uravneniya i volny v sluchaino-neodnorodnykh sredakh* (in Russian) (Stochastic Equations and Waves in Randomly Inhomogeneous Media) Moscow, Nauka, 1980, 336 p.
83. Montroll, E. W., *J. Chem. Phys.*, 1946, vol. 14, no. 3, pp. 202–211.
84. Christian, J. W., *The Theory of Transformations in Metals and Alloys*, Pt. I, Pergamon Press, Oxford, 1975, 586 p.
85. Bullough, R., Newman, R., *Progr. Phys.*, 1970, vol. 33, no. 2, p. 101.
86. Berlin, Yu. A., Gordier, P., Delaire, J. A., *J. Chem. Phys.*, 1980, vol. 73, no. 9, pp. 4616–4627.
87. Hummel, A., *Adv. Rad. Chem.*, 1974, vol. 4, no. 1, pp. 1–102.
88. Butter, W. H., Fromhold, A. T., *J. Phys. Chem. Solids.*, 1974, vol. 35, no. 9, pp. 1099–1113.
89. Siger, A., *Phys. Lett.*, 1972, vol. 38-A, no. 1, pp. 45–46.
90. Hong, K. M., Noolandi, J., *J. Chem. Phys.*, 1978, vol. 68, no. II, pp. 5163–5172.
91. Fuks, N. A., *Mekhanika aerozolei* (in Russian) (The Mechanics of Aerosols) Moscow, USSR Acad. Sci. Publ. House, 1955, 358 p.
92. Debye, P., *Trans. Electrochem. Soc.*, 1942, vol. 32, pp. 265–272.
93. Onsager, L., *Phys. Rev.*, 1938, vol. 54, pp. 554–557.
94. Nagri, R. K., Waldenstrom, S., Mork, J., *J. Chem. Phys.*, 1979, vol. 71, no. I, p. 73–80.
95. Belyi, A. A., *ZhFKh* (in Russian) (J. Phys. Chem.), 1984, vol. 58, no. 9, pp. 2189–2193.
96. Umberger, J. Q., La Mer, V. K., *J. Amer. Chem. Soc.*, 1945, vol. 67, no. 7, pp. 1099–1109.
97. Allen, A. O., Holroyd, R. A., *J. Phys. Chem.*, 1974, vol. 78, no. 8, pp. 796–803.
98. Entelis, S. G., Tiger, R. P., *Kinetika reaktsiyi v zhidkoi faze* (in Russian) (Reaction Kinetics in Liquid Phase) Moscow, Khimiya, 1973, 416 p.
99. Bass, L., Greenhalgh, W. J., *Trans. Faraday Soc.*, 1966, vol. 62, no. 3, pp. 715–722.
100. Gősele U., *Chem. Phys. Lett.*, 1980, vol. 69, no. 2, pp. 332–335.
101. Tyutnev, A. P., Sichkarev, V. P., Vannikov, A. V., *Usp. Khim.* (in Russian) (Adv. in Chem.), 1981, vol. 50, no. 6, pp. 977–1106.
102. Maggee, J. L., Taylor, A. B., *J. Chem. Phys.*, 1972, vol. 56, no. 6, pp. 3061–3066.
103. Rice, S. A., et al, *Ibid.*, 1979, vol. 79, no. 9, pp. 4001–4007.

104. Sibani, P., Pedersen, J. B., *Ibid.*, 1981, vol. 74, no. 12, pp. 6934–6942.
105. Pedersen, J. B., Sibani, P., *Ibid.*, 1981, vol. 75, no. II, pp. 5368–5372.
106. Ovchinnikov, A. A., Burlatsky, S. F., *Chem. Phys.*, 1981, vol. 63, no. I, pp. 67–75.
107. Williams, F., *J. Am. Chem. Soc.*, 1964, vol. 86, no. 19, pp. 3954–3957.
108. Hummel, A., Infelta, P. P., *Chem. Phys. Lett.*, 1974, vol. 24, no. 4, pp. 559–562.
109. Schulten, Z., Schulten, K., *J. Chem. Phys.*, 1977, vol. 66, no. 10, pp. 4616–4634.
110. Hynes, J. T., Kapral, R., Torrie, G. M., *Ibid.*, 1980, vol. 72, no. I, pp. 177–188.
111. Delaire, J. A., Croc, E., Cordier, P., *Ibid.*, 1981, vol. 85, no. II, pp. 1549–1555.
112. Rice, S. A., Baird, J. K., *Ibid.*, 1978, vol. 69, no. 5, pp. 1989–1995.
113. Zamraev, K. I., Khairutdinov, R. F., *Usp. khim.* (in Russian) (Adv. in Chem.), 1978, vol. 47, no. 6, pp. 992–1007.
114. Burshtein, A. I., *Ibid.*, no. 2, p. 212–234.
115. Tunitskii, N. N., Bagdasaryan, H. S., *Opt. i spektr.* (in Russian) (Optics and Spectroscopy), 1963, vol. 15, no. I, pp. 100–106.
116. Gradshtein, I. S., Ryzhik, I. M., *Tablitsy integralov, summ, ryadov i proizvedenii* (in Russian) (Tables of Integrals, Sums, Series, and Products) Moscow, GMFML, 1963, 1100 p.
117. Doktorov, A. B., Burshtein, A. I., *ZhETF* (JETP), 1975, vol. 68, no. 4, pp. 1349–1362.
118. Lawson, C. M., Freed, E. E., Powell, R. C., *J. Chem. Phys.*, 1982, vol. 76, no. 8, pp. 4171–4177.
119. Heisel F., Mieche, J. A., *Ibid.*, vol. 77, no. 5, pp. 2558–2569.
120. Antipenko, B. M., Ermolaev, V. L., *Opt. i spektr.* (Opt. and Spectroscopy), 1970, vol. 28, no. 5, pp. 931–933; vol. 29, no. I, pp. 90–92.
121. Ermolaev, V. L., Grusdkov, V. P., Tachin, V. S., *Izv. AN SSSR. Ser. fiz.* (USSR Acad. Sci Proceedings, Ser. Phys.), 1972, vol, 36, no. 5, pp. 984–987.
122. Bodunov, E. N., *Opt. i spektr.* (in Russian) (Opt. and Spectroscopy), 1973, vol. 34, no. 3, pp. 490–496.
123. Bodunov, E. N., *Izv. AN SSSR. Ser. fiz.* (USSR Acad. Sci. Proc., Ser. Phys.), 1972, vol. 36, no. 5, pp. 796–799.
124. Doktorov, A. B., Kipriyanov, A. A., Burshtein, A. I., *Opt. i spektr* (in Russian) (Opt. and Spectroscopy), 1978, vol. 45, no. 3, pp. 497–504; vol. 45, no. 4, pp. 684–690.
125. Steen, H. B., Strand, O., *J. Chem. Phys.*, 1974, vol. 60, no. 12, pp. 5043–5049.
126. Funabashi, K., Hebert, C., Maggee, J. L., *Ibid.*, 1971, vol. 75, no. 21, pp. 3221–3229.
127. Bagdasaryan, N. S., *Dvuknkvantovaya fotokhimiya* (in Russian) (Two-Quantum Photochemistry) Moscow, Nauka, 1976, 127 p.
128. Tachiya, M., Mozumder, A., *Chem. Phys. Lett.*, 1975, vol. 34, no. I, pp. 77–79.

129. Zamraev, K. I., *Khim. fizika* (in Russian) (Chem. Physics), 1982. no. 2, pp. 156–176.
130. Berlin, Yu. A., *Dokl. Akad, Nauk SSSR* (Reports of USSR Acad. Sci.), 1975, vol. 223, no. 3, pp. 625–628; no. 6, pp. 1387–1390.
131. Berlin, Yu. A., *Ibid.*, 1976, vol. 228, no. 3, pp. 628–631.
132. Pilling, M. J., Rice, S. A., *J. Phys. Chem.*, 1975, vol. 79, no. 26, pp. 3075–3038.
133. Pilling, M. J., Rice, S. A., *J. Chem. Soc. Faraday Trans.*, Pt. 2, 1975, vol. 71, no. 9, pp. 1563–1571.
134. Butler, P. R., Pilling, M. J., *Discus. Faraday Soc.*, 1977, vol. 63, pp. 38–46.
135. Kotomin, E., Fabrikant, I., *J. Phys. C. Solid State Phys.*, 1977, vol. 10, no. 24, pp. 4931–4937.
136. Berlin, Yu. A., Khairutdinov, R. F., *Khimiya vysokikh energiyi* (The Chemistry of High Energies), 1978, vol. 12, no. I, pp. 3–7.
137. *Handbook of Mathematical Functions with Formulas, Graphs and Mathematical Tables,* Ed. by M. Abramovitz, I. A. Stegun, Natl. Bureau Stand., Appl. Math. Series, 55, 1964.
138. Caldin, E. F., *Fast Reactions in Solution,* Blackwell Scientific Publications, Oxford, 1964.
139. Denisov, E. T., *Usp. khim.* (in Russian) (Adv. in Chem.), 1970, vol. 39, no. I, pp. 62–93.
140. Klafter, J., Zumonften, G., Blumen, A., *J. Phys. Lett.*, 1984, vol. 45, no. 2, p. 49–56.
141. Monchick, Z., Samuel, A. M., Magee, J. L., *J. Chem. Phys.*, 1957, vol. 26, no. 4, pp. 935–941.
142. Paul, R., *Ibid.*, 1968, vol. 49, no. 6, pp. 2806–2815.
143. Shekhtman, V. Z., *Opt. i spektr.* (in Russian) (Opt. and Spectroscopy), 1972, vol. 33, no. 2, pp. 284–291.
144. Doktorov, A. V., Kipriyanov, A. A., Burshtein, A. I., *ZhETF* (JETP), 1978, vol. 74, no. 3, pp. 1184–1198.
145. Kotomin, E. A., Kuzovkov, V. N., *Teoret. i eksper. khim.* (Theor. and Exper. Chem.), 1982, no. 3, pp. 274–279.
146. Kalnin, Yu. H., Pirogov, F. V., *Phys. Status Solidi. B.*, 1977, vol. 84, no. 2, pp. 521–527.
147. Belyi, A. A., Ovchinnikov, A. A., *Teor. i eksper. khim.* (Theor. and Exper. Chem.), 1967, vol. 3, no. 4, pp. 451–457.
148. Belyi, A. A., Ovchinnikov, A. A., *Zhur. Fiz. Khim.* (J. of Chem. Phys.), 1970, vol. 44, no. II, pp. 2824–2829.
149. Antonov-Romanovskii, V. V., *Kinetica fotolyuminestsentsii kristallofosforov* (in Russian) (The Kinetics of Crystallophosphors Photoluminescence) Moscow, Nauka, 1966, 324 p.
150. Kofman, A. G., Burshtein, A. I., *Chem. Phys.*, 1978, vol. 27, no. 2, pp. 217–228.
151. Gősele, U., Seeger, A., *Phil. Mag.*, 1976, vol. 34, no. 2, pp. 177–193.
152. Aityan, S. H., *Biologicheskie membrany* (in Russian) (Biological Membranes), 1984, vol. I, no. 5, pp. 524–530.
153. Markin, V. S., Chizmadzhev, Yu. A., *Indutsirovannyi ionnyi transport* (in Russian) (Induced Ion Transport) Moscow, Nauka, 1974, 252 p.

154. Ovchinnikov, A. A., Belyi, A. A., *Teoret. i eksper. khim.* (Theoret. and Exper. Chem.), 1966, no. 3, pp. 538–542.
155. Ovchinnikov, A. A., Zeldovich, Ya. B., *Chem. Phys.*, 1978, vol. 28, no. 1,2, pp. 215–218.
156. Mikhailov, A. I., et al, *Fiz. Tver. Tela* (in Russian) (Solid State Phys.), 1972, vol. 14, no. 4, pp. 1172–1179.
157. Garbovitskaya, T. I., Belevskii, V. N., Bugaenko, L. T., *Vestnik MGU. Ser. khim.* (MGU Bulletin, Chem. Series), 1973, vol. 14, no. 2, pp. 163–166.
158. Lebedev, Ya. S., Burshtein, A. I., *Chem. Phys.*, 1976, vol. 12, no. 3, pp. 259–265.
159. Yakimchenko, O. E., Lebedev, Ya. S., *Uspekh. khim.* (Adv. in Chem.), 1978, vol. 47, no. 6, v.1018–1047.
160. Khudyakov, I. V., Levin, P. P., Kuzmin, V. A., *Uspekhi khim.*, (Adv. in Chem.), 1980, vol. 49, no. 10, pp. 1990–2031.
161. Emanuel, N. M., Roginskii, V. A., Buchachenko, A. L., *Ibid.*, 1982, vol. 51, no. 3, pp. 361–393.
162. Simpson, H. M., Sosin, A., *Radiation Effects*, 1970, vol. 3, pp. 1–21.
163. Belyi, A. A., Ovchinnikov, A. A., *Khimiya yysokikh energii* (High Energy Chemistry), 1967, vol. I, no. 4, pp. 346–351.
164. Belyi, A. A., Ovchinnikov, A. A., *Ibid.*, 1971, vol. 5, pp. 331–336.
165. Roginskii, V. A., Kotov, B. V., *Ibid.*, 1967, vol. I, no. 4, pp. 291–308.
166. Huber, D. L., *Phys. Rev. B.*, 1979, vol. 20, no. 12, pp. 5333–5338.
167. Mansur, L. K., *J. Nucl. Mater.*, 1979, vol. 83, no. I, pp. 109–121.
168. Felderhof, B. U., Deutch, J. M., Titulaer, U. M., *J. Chem. Phys.*, 1976, vol. 64, no. II, pp. 4551–4559; 1982, vol. 76, no. 8, pp. 4178–4184.
169. Ham, F. S., *J. Chem. Phys. Solid.*, 1958, vol. 6, pp. 335–351.
170. Carlaw, H. S., Jaeger, J. C., *Conduction of Heat in Solids.*, Oxford, Clarendon Press.
171. Ziman, J. M., *Principles of the Theory of Solids,* Cambridge University Press, 1964.
172. Agranovich, V. M., Galanin, M. D., *Perenos energii elektronnogo vozbuzhdeniya v kondensirovannykh sredakh* (in Russian) (The Transfer of Electron Excitation Energy in Condensed Media). Moscow, Nauka, 1978, 383 p.
173. Ham, F. S., *J. Appl. Phys.*, 1959, vol. 30, no. 6, pp. 915–926.
174. Kaminskii, V. A., Okunev, B. N., Ovchinnikov, A. A., *Dokl. Akad. Nauk SSSR* (USSR Acad. Sci. Reports), 1980, vol. 251, no. 3, pp. 636–639.
175. Muthukumar, M., *J Chem. Phys.*, 1982, vol. 76, no. 5, pp. 2667–2671.
176. Braisfold, A. D., *J. Nucl. Mater.*, 1976, vol. 60, no. 3, pp. 257–278.
177. Freed, K. F., Metiu, H., *J. Chem. Phys.*, 1978, vol. 68, no. 10, pp. 4604–4611.

178. Elliott, R. L., Krumhansl, J. A., Leath, P. L., *Rev. Modern Physics,* 1974, vol. 46, p. 465.
179. Braisfold, A. D., Bullough, R., *J. Nucl. Mater.*, 1972, vol. 44, no. 2, pp. 121–125.
180. Braisfold, A. D., Bullough, R., Hayns, M. R., *J. Nucl. Mater.*, 1976, vol. 60, no. 3, pp. 246–256.
181. Tabolt, D. R. S., Willis, J. R., *Proc. Royal Soc. London,* 1980, vol. A 370.
182. Woo, C. H., *J. Nucl. Mater.*, 1981, vol. 98, no. 3, pp. 279–294.
183. Burlatsky, S. F., *Avtoreferat kand. dis.* (in Russian) (The Author's Abstract of a Master's Thesis), Moscow, Inst. of Chem. Phys. USSR Acad. Sci., 1979.
184. Burlatsky, S. F., *Dokl. Akad. Nauk SSSR* (Reports of USSR Acad. Sci.), 1979, vol. 247, no. 2, pp. 373–375.
185. Wise, H., Ablow, C. M., *J. Chem. Phys.*, 1958, vol. 29, no. 3, pp. 634–639.
186. Danckwerts, P. V., *Gas-liquid Reactions. Second Edition,* McGraw-Hill Book Company, N.Y., 1970.
187. Shurr, J. M., *Biophys. J.*, 1970, vol. 10, pp. 700–716.
188. Prager, S., Frish, H. L., *J. Chem. Phys.*, 1975, vol. 62, no. I, p. 88–91.
189. Landau, L. D., Lifshits, E. M., *Statisticheskaya fizika* (Statistical Physics). Moscow, Nauka, 1976, pt. I, 584 p.
190. Zeldovich, Ya. B., *Electrokhimiya* (Electrochemistry), 1977, vol. 13, no. 5, pp. 677–679.
191. Belyi, A. A., Ovchinnikov, A. A., Timashev, S. F., *Teoret. i eksper. khim.* (Theoret. and Experim. Chem.), 1982, vol. 18, no. 3, p. 269–274.
192. Fok, V. A., *Raboty po kvantovoi teorii polya* (in Russian) (Works on Field Quantum Theory). Leningrad, LGU Publ. House, 1957, 158 p.
193. Prigogine, I., *Introduction to the Thermodynamics of Irreversible Processes,* N.Y., Wiley, 1961, 119 p.
194. Abrikosov, A. A., Gor'kov, L. P., Dzyaloshinsky, I. E., *Metody kvantovoi teorii polya v statisticheskoi fizike* (in Russian) (Methods of Field Quantum Theory in Statistic of Physics) Moscow, Fizmatgiz, 1962, 443 p.
195. Bogolyubov, N. N., *Izv. Akad. Nauk SSSR, Ser. fiz.* (USSR Acad. Sci. Bulletin, Ser. Phys.), 1947, vol. II, no. I, p. 77–90.
196. Van Kampen, N., Oppenheim, I., *J. Math. Phys.*, 1972, vol. 13, no. 6, pp. 842–849.
197. Bashkirov, A. G., *V kn; Molekulyarnaya gazodinamika* (in Russian) (In: Molecular Gas Dynamics). Moscow, Nauka, 1982, pp. 107–120.
198. Waldman, L., *Z. Naturforsch.*, 1967, volume 22A, no. 8, pp. 1269–1280.
199. Bedeaux, D., Albano, A., Mazur, P., *Physica,* 1976, vol. 82A, no. 3, pp. 438–462.
200. Vrunov, P. A., Kaminskii, V. A., Timashev, S. F., *V kn: Materialy V Vsesoyuznogo simposiuma po mekhanoemissii i mekhanokhimii tverdykh tel.* (in Russian) (Abstracts of the *V* All-Union Symposium of Mechanical Emission and Mechanical Chemistry of Solids, Tallinn, October 1975). Tallinn, 1977, pp. 232–235.

201. Kaminskii, V. A., Vrunov, P. A., Timashev, S. F., *Kolloid zhurn.* (Colloid Journal), 1977, vol. 39, no. 3, pp. 692–697.
202. Burshtein, A. I., Kofman, A. G., *Opt. i spektroskopiya* (Opt. and Spectr.), 1976, vol. 40, no. 2, pp. 304–310.
203. Belyi, A. A., et al, *Dokl. Akad. Nauk SSSR* (USSR Acad. Sci. Reports), 1978, vol. 242, no. 6, pp. 856–859.
204. Merzhanov, A. G., Bloshenko, V. N., *Ibid.*, no. 5, pp. 1118–1121.
205. Belyi, A. A., Ovchinnikov, A. A., Timashev, S. F., *ZhTF* (in Russian), (JTP), 1979, vol. 49, no. 9, pp. 1987–1992.
206. Monchick, L., *J. Chem. Phys.*, 1975, vol. 62, no. 5, pp. 1907–1912.
207. Northrup, S., Hynes, J. T., *Ibid.*, 1979, vol. 71, no. 2, pp. 871–893.
208. Brown, L. C., *J. Appl. Phys.*, 1976, vol. 47, no. 2, pp. 449–458.
209. Northrup, S., Hynes, J. T., *J. Stat. Phys.*, 1978, vol. 18, no. I, pp. 91–105.
210. Fikus, J. H., Johnson, R. A., *Radiation Effects,* 1979, vol. 40, pp. 63–70.
211. Ovchinnikov, A. A., Belyi, A. A., Timashev, S. F., *Kinetika i mekhanizm khimicheskikh reaktsii v tverdom tele* (in Russian) (Kinetics and Mechanism of Chemical Reactions in Solids) Preprint. Chernogolovka, IPC USSR Acad. Sci., 1981, pp. 66–79.
212. Krykin, M. A., Timashev, S. F., Lomakin, V. V., *Dokl. Akad. Nauk SSSR* (USSR Acad. Sci. Reports), vol. 270, no. 5, pp. 1148–1153.
213. Lifshits, A. I., Metter, S. F., Samartsev, A. A., *ZhTF* (JTP), 1976, vol. 46, no. 7, pp. 1490–1500.
214. Industrial Processing With Membranes. Ed. R. E. Lacey and S. Loeb, Wiley-Interscience, Inc., N.Y., 1972.
215. Nikolaev, N. I., *Diffuziya v membranakh* (in Russian) (Diffusion in Membranes), Moscow, Khimiya, 1980, 232 p.
216. Hwang, S.-T., Kammermeyer, K., *Membranes in Separations*. N.Y., Wiley, 1975, 559 p.
217. Reitlinger, S. A., *Pronitsaemost polimernykh membran* (in Russian) (The Permeability of Polymer Membranes). Moscow, Khimiya, 1974, 269 p.
218. Gnusin, N. P., Grebenyuk, V. D., Pevnitskaya, M. V., *Electrokhimiya ionitov* (in Russian) (The Electrochemistry of Ion Exchangers) Moscow, Nauka, 1972, 200 p.
219. Grebenyuk, V. D., *Electrodializ* (in Russian) (Electrodialysis) Kiev, Naukova dumka, 1976, 147 p.
220. Dytnerskii, Yu. I., *Obratnyi osmos i ultrafiltratsiya* (in Russian) (The Reverse Ososis and Ultrafiltration). Moscow, Khimiya, 1978, 352 p.
221. Dytnerskii, Yu. I., *Membrannye protsessy razdeleniya zhidkikh smesei* (in Russian) (Membrane Processes of Liquid Mixture Division). Moscow, Khimiya, 1975, 232 p.
222. Yasminov, A. A., et al, *Obrabotka vody obratnym osmosom i ultrafiltratsiei* (in Russian) (Water Treatment by the Reverse Osmosis and Ultrafiltration). Moscow, Stroiizdat, 1978, 121 p.
223. Dubyaga, V. P., Perepechkin, L. P., et al, *Polimernye membrany* (in Russian) (Polymer Membranes). Moscow, Khimiya, 1981, 232 p.

224. *Ion Exchange Membranes*. Ed. D. S. Flett, London, Ellis Horwood Ltd., 1983, 210 p.
225. Timashev, S. F., Gladkikh, S. H., *Dokl. Akad. Nauk SSSR* (USSR Acad. Sci. Reports), 1982, vol. 262, no. 3, pp. 656–661.
226. Timashev, S. F., *ZhFkh* (J. Phys. Chem.), 1982, vol. 56, no. 7, pp. 1739–1742.
227. Timashev, S. F., *ZhFkh* (J. Phys. Chem.), 1983, vol. 57, no. I, pp. 68–74.
228. Timashev, S. F., Popkov, Yu. M., Slesarev, Yu, V., *Teoret. osnovy khim. tekhnologii* (Theoret. Principles of Chem. Technology) 1984, vol. 18, no. 2, pp. 165–170.
229. Timashev, S. F., Kirganova, E. V., *Electrokhimiya* (Electrochemistry), 1981, vol. 17, no. 3, pp. 440–443.
230. Kirganova, E. V., Timashev, S. F., Popkov, Yu. M., *Ibid.*, 1983, vol. 19, no. 7, pp. 973–980.
231. Wolf, W. E., Arvanitis, S., Simon, W., *Chimia,* 1979, vol. 33, pp. 452–460.
232. Schults, J. S., Goddard, J. D., Sucheo, S. R., *AlChE J.*, 1974, vol. 20, pp. 417, 625.
233. Timashev, S. F., *Zashchita metallov* (Protection of Metals), 1980, vol. 16, no. 2, pp. 176–180.
234. Fetter, K., *Electrokhimicheskaya kinetika* (in Russian) (Electrochemical Kinetics). Moscow, Khimiya, 1967, pp. 856.
235. Timashev, S. F., *ZhFkh* (J. Phys. Chem.), 1982, vol. 56, no. 7, pp. 1736–1742.
236. Timashev, S. F., *ZhFkh* (J. Phys. Chem.), 1983, vol. 57, no. I, pp. 68–74.
237. Sata, T., et al, J. Polymer Sci., Polymer Chem. Ed., 1979, vol. 17, pp. 2071–2075.
238. Kotov, V. V., Zolotareva, R. I., Minakova, T. P., *Zh. prikl. khimii* (J. Appl. Chem.), 1981, vol. 54, pp. 1433–1436.
239. Kleijn, W. B., Bruner, L. J., *J. Theoretical Biol.*, 1983, vol. 100, pp. 135–152.
240. Timashev, S. F., *Biofizika* (Biophysics), 1981, vol. 26, no. 4, pp. 642–646.
241. Mauro, A., *Biophys.*, 1962, vol. 2, pp. 179–198.
242. Kirganova, E. V., Timashev, S. F., Popkov, Yu. M., *V kn: VI Vsesoyuznaya konf. po elektrokhimii. Tez. dokl.* (in Russian) (In: VI All-Union Conf. on Electrochemistry. Abst. Rep.), June 21–25, 1982, Moscow, VINITI, 1982, vol. 2, p. 161.
243. Greben, V. P., et al, *ZhFkh* (J. Phys. Chem.), 1978, vol. 52, no. 10, pp. 2641–2645.
244. Gagarin, S. G., Chuvylkin, N. S., *ZhFkh* (J. Phys. Chem.), 1981, vol. 55, no. 12, pp. 3089–3093.
245. Bobreshovia, O. V., Timashev, C. F., *Electrokhimiya* (Electrochemistry), 1985, vol. 21, no. II.
246. Nikonenko, N. P., *Ibid.*, 1979, vol. 15, no. 10, pp. 1494–1502.
247. Falk, M., *Can J. Chem.*, 1980, vol. 58, pp. 1495–1503.
248. Gladkikh, S. N., et al, *Vysokomol. soed.* (High-Molecul. Compounds) 1984, vol. 26B, no. I, pp. 20–23.
249. Gierke, T. D., Munh, G. F., Wilson, F. C., *J. Polymer Sci.*, *Polym. Phys. Ed.*, 1981, vol. 19, pp. 1687–1704.

250. Ozerin, A. N., et al., *Vysokomol. soed.* (High-Molecul. Comp.) 1986, vol. 28A, no. 3.
251. Heitner-Wirguin, C., et al., *Polymer,* 1980, vol. 21, pp. 1327–1329.
252. Volkov, V. I., et al., *Khim. fizika* (Chem. Physics), 1983, vol. 2, no. I, pp. 49–53.
253. Volkov, V. I., et al., *Ibid.*, 1985, vol. 4, no. 5.
254. Volkov, V. I., et al., *Dokl, Akad. Nauk SSSR* (Reports of USSR Acad. Sci.), 1984, vol. 276, no. 2, pp. 395–400.
255. Volfkovich, Yu, M., et al., *Electrokhimiya* (Electrochemistry), 1984, vol. 20, no. 5, pp. 656–664.
256. Romashin, O. P., Fioshin, M. M., Erenburg, R. G., *Ibid.*, 1979, vol. 15, no. 5, pp. 653–659.
257. Timashev, S. F., Gladkikh, S. N., *Dokl. Akad. Nauk SSSR* (Reports of USSR Acad. Sci.), 1982, vol. 262, no. 3, pp. 656–661.
258. Shklovskii, B. I., Efros, A. L., *Usp. fiz. nauk* (Adv. Phys. Science), 1975, vol. 117, no. 3, pp. 401–435.
259. Skal, A. S., Shklovskii, B. I., *Fizika i tekhnika polyprovodnikov,* (The Physics and Technology of Semiconductors). 1974, vol. 8, no. 4, pp. 1586–1592.
260. Yeager, M. Z., Kipling, B., Dotson, R. L., *J. Electrochem. Soc.*, 1980, vol. 127, no. 2, pp. 303–306.
261. Nicholson D., Parsonage, N. G., *In: Computer Simulation and the Structural Mechanism of Absorbtion.* London, Academic Press, 1982, 369 p.
262. Glueckanf, E., Russel, P. J., *Desalination,* 1970, vol. 8, no. 2, pp. 351–257.
263. Dukhin, S. S., Gaevskii, A. Yu., Yaroschuk, A. E., *Khimiya i technologiya vody* (Water Chemistry and Technology), 1983, vol. 5, no. I, pp. 13–21.
264. Heyde, M. E., Anderson, G. E., *J. Colloid Interface Sci.,* 1975, vol. 50, no. 2, pp. 467–488.
265. Deryagin, B. V., Churaev, N. V., *Progr. Surface and Membrane Sci.,* 1981, vol. 14, pp. 69–130.
266. Deryagin, B. V., Churaev, N. V., *Kolloid. zhurn.* (in Russian) (Colloid. Jour.), 1980, vol. 42, no. 5, pp. 842–851.
267. Barer, S. S., et al, *Dokl. Akad. Nauk SSSR* (in Russian) (Reports of USSR Acad. Sci.), 1977, vol. 235, no. 3, pp. 601–604.
268. Mitchell, D. J., Ninharm, B. W., Pailthorpe, B. A., *J. Colloid. Interface Sci.,* 1978, vol. 64, no. I, pp. 194–198.
269. Chan, D. Y. C., Mitchell, D. J., Ninhan, B. W., *Mol. Phys.,* 1978, vol. 3, no. 7, pp. 1669–1675.
270. Israelashvili, J. N., Adams, G. E., *J. Chem. Soc. Faraday Trans. I.,* 1978, vol. 74, no. 6, pp. 975–983.
271. Israelashvili, J. N., *Discus. Farad. Chem. Soc.*, 1978, no. 65, pp. 20–24.
272. Kiseleva, O. A., et al, *Kolloid. zhurn.* (in Russian) (Colloid. Journ.), 1979, vol. 41, no. 2, pp. 245–249.
273. Deryagin, B. V., et al, *Zhurn. tekhin. fiziki* (in Russian) (J. Tech. Physics), 1959, vol. 27, no. 9, pp. 1976–1981.
274. Toprak, C., Agar, J. N., Falk, M., *Trans. Faraday Soc.*, 1979, vol. 75, no. 3, pp. 803–815.

275. Scherer, J. R., Bailey, G. F., *J. Membrane Sci.*, 1983, vol. 13, no. I, pp. 29–52.
276. Kurihara M., Himeshima, Y., Veumure, T., *The 1987 Int. Congress on Membrane Processes,* Tokyo, 1987, 428 p.
277. Timashev, S. F., *Uspekhi Khim.* (in Russian) (Adv. in Chem.), 1988, vol. 57, no. 6, pp. 876–902.
278. Kirsh, Yu, E., Popkov, Yu. M., *Ibid.*, 1988, vol. 57, no. 6, pp. 1001–1009.
279. Neymerk, A. V., *Matematicheskie metody opisaniya gornykh porod i rascheta ikh effectivnykh svoistv* (in Russian) (Mathematical Method of Describing Rock and Calculating its Properties). Moscow, Nauka, 1986, p. 80.
280. Higashi, N., Kunitaka, K., Kajiuma, T., *The 1987 Int. Congress on Membrane and Membrane Processes,* Tokyo, Japan, 1987, 675 p.
281. Dugaryan, S. G., Yampolskii, Yu. P., *Neftekhimiya* (in Russian) (Petroleum Chemistry), 1983, vol. 23, no. 5, pp. 579–595.
282. Henis, J. M. S., Tripodi, M. K., *Separation Sci. and Technol.*, 1980, vol. 15(4), no. 5, pp. 1059–1068.
283. Lifshits, A. I., *ZhTF* (in Russian) (JTP), 1976, vol. 56, pp. 328–338.
284. Lifshits, A. I., Samartsev, A. A., *Ibid.*, 1979, vol. 49, no. II, pp. 2433–2436.
285. Krykin, M. A., Timashev, S. F., *Fizika tverdogo tela* (in Russian) (Physics of Solids), 1985, vol. 27, no. 9, pp. 2549–2556.
286. Valetskii, P. M., Storozhuk, I. P., *Usp. khim.* (in Russian) (Adv. in Chem.), 1979, vol. 48, pp. 75–114.
287. Teplyakov, V. V., et al, *Plastmassy* (in Russian) (Plastics), 1978, no. 5, p. 49–51.
288. Abdyukhanov, V. A., Nikolaev, N. I., *In: Abstr. of All-Union Conf. on Membrane Methods of Mixtures Division* (in Russian), Vladimir, April 12–14, 1977), Vladimir, VNIISS, 1977, 54 p.
289. Lifshits, A. I., Nofkin, M. E., Yakovlev, S. V., *Pis'ma v ZhTF,* (in Russian) (JTP Letters), 1978, vol. 4, 8th Ed., pp. 476–478.
290. Raigorodskii, I. M., et al, *Vysokomol. soed. Kratkie soobscheniya,* (in Russian) (High-molec. Compounds, Bried Reports), 1984, vol. XXVI (Ser B), no. I, pp. 25–28.
291. Sakharov, S. E., *Khim. Fizika* (in Russian) (Chemical Physics), 1986, vol. 5, no. 2, pp. 272.

SUBJECT INDEX